40 YEARS OF SHS: A LUCKY STAR OF A SCIENTIFIC DISCOVERY
A Presentation with Elements of a Scientific Lecture

By

A.G. Merzhanov

*Institute of Structural Mackrokinetics and
Materials Science (ISMAN)
Russia*

CONTENTS

FOREWORD

Dear Readers,

You are holding an unusual book in your hand. It represents a view of one of the authors of the scientific discovery of "Solid flame phenomenon" academician A.G. Merzhanov, who is a recognized leader of the scientific-and-technical area of "Self-propagating high-temperature synthesis" (SHS) all over the world, on the forty-year path from the scientific discovery to science-intensive production.

It is neither a book of memoirs nor a scientific monograph; it's "a presentation with elements of a scientific lecture" as the author determined it. The book is an expanded version of the lecture delivered by A.G. Merzhanov at the International conference dedicated to the 40-th anniversary of SHS. He is telling about people, scientific discoveries and important events rather emotionally and brightly. Besides there is a set of illustrations: diagrams, tables, schemes, photos which demonstrate the most remarkable (from the author's viewpoint) fragments of the variegated picture of the SHS development. The author is paying much attention to a little-known initial stage of the investigations in the former USSR. At the same time a modern situation in the SHS area is being considered and the most promising tasks for future development are being defined.

Perhaps, you won't agree with the conclusions of the author. Probably, you consider some other events and results to be more significant. It's a normal situation for such an alive and growing area as SHS. The author characterizes this work as "a subjective opinion of a person who has been working within the direction actively for 40 years". But we are sure that the open-hearted, well thought-out position of the Father of SHS, his keenness and energy won't leave the readers of this book indifferent.

A.S. Rogachev,

A.S. Mukasyan

PREFACE

I have prepared this paper because before the 40-th anniversary of SHS I wanted to share my knowledge, impressions and ideas with you and tell you about the origin of the investigation in this field, about the Pioneers of this direction, their achievements and interesting result.

The task appeared to be rather complicated because I had to expound the material thoroughly without any faults. It was difficult, and I know that some events were mixed "in time and space" and I paid more attention to the initial stage of our activity. But it was the most important period which was hardly described in literature.

Besides, I know the work carried out in our Institute much better than that made by our colleagues from other organizations. That is why I am sorry, but there are more examples of our achievements. Submitting the paper to your judgment, I'd like to underline that I consider it as an essay. I am trying to express a subjective opinion of a person who has been working within the direction actively for 40 years.

Also I'd like to express my gratitude to all my colleagues who helped me in preparing the manuscript.

A.G. Merzhanov
Institute of Structural Mackrokinetics and
Materials Science (ISMAN)
Russia

INTRODUCTION

In 1967 a group of young scientists at the Branch of the Institute of Chemical Physics of the USSR Academy of Sciences was lucky to make a scientific discovery [1,3]. At first it seemed to be plain but later appeared to be rather significant and fruitful. While investigating the combustion mechanism of solid propellants and searching for experimental gasless combustion models, they discovered a new type of unknown combustion processes in which all the substances (initial components, intermediate and final products) were solid. Such a process is shown in Fig. **0.1**. It is propagation of a high-temperature chemical reaction in the powder mixture of tantalum and carbon. It is obvious that the process can occur when the maximum temperature in the combustion wave (it is called a combustion temperature) is less than the eutectic point known from the state diagram of the system under study. The comparison of these values for Ta-C proves that this requirement is satisfied. The typical picture of such solid flame combustion (SFC) is presented in Fig. **0.2**. After the ignition the combustion front which serves as a boundary between the initial mixture and the hot but solid product propagates along the mixture and the burnt-up part of the sample gets cold.

The experimental sample is made as a compact cylindrical pellet to be ignited at the end. It is a conventional model used in experiments and theoretical investigations.

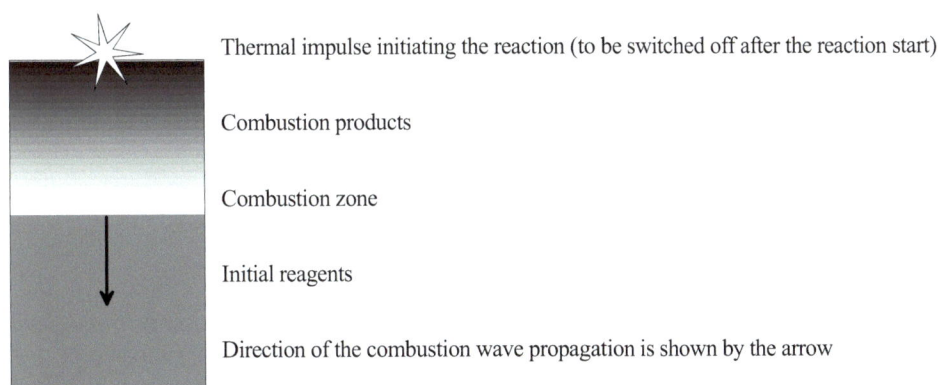

Thermal impulse initiating the reaction (to be switched off after the reaction start)

Combustion products

Combustion zone

Initial reagents

Direction of the combustion wave propagation is shown by the arrow

Fig. (0.1). Patterns of the "solid flame" process.

By the way, the picture of solid flame combustion which was made by the authors for illustrating the new process has become very popular and is presented in many papers. Why was the process recognized as a scientific discovery? In order to answer this question I'd like to quote some words from my monograph "Solid flame combustion": "The investigation of SFC is based on the scientific discovery made in the USSR Academy of Sciences in 1967 and registered as "a phenomenon of autowave localization of autoretarding solid-phase reactions". This name should be explained. Solid substances react with each other in the contact point. The solid substances appearing in such a case form a barrier separating the reagents. The further interaction of the reagents occurs due to their diffusion through the barrier layer with the velocity decreasing with time. Thus, the reaction retards itself – it explains the name "an autoretarding reaction". These slow autoretarding reactions initiated locally can be concentrated in the zone and occur in such fast modes as combustion wave propagation".

Fig. (0.2): Combustion front propagation.

But I'd like to dwell on the beginning of the work. We liked the phenomenon of solid flame as a discovery. We realized that it could be developed in three ways, and being lost in our day-dreams we made the following scheme of its further development (Fig. **0.3**) without believing in its reality.

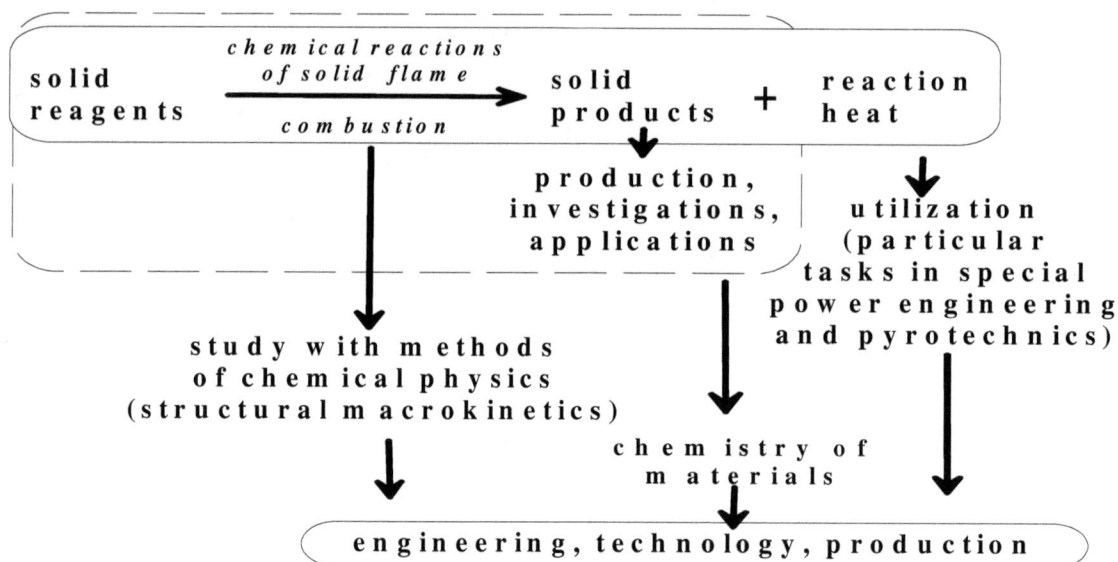

Fig. (0.3): Solid flame combustion, self-propagating high temperature synthesis and macroscopic kinetics. SHS=obtaining of refractory compounds during gasless combustion.

The easiest way was to investigate the process. The team, in charge of which I was lucky to be, included highly qualified specialists in the field of macroscopic kinetics, combustion theory and practice. This way was clear to be the best as it provided a lot of possibilities for obtaining new results interesting for the combustion science. However, we realized that the subject of investigation was not only the process but the product as well, and if the combustion product was of interest for our practice, the solid flame combustion could become a synthesis method. It was a complicated task for our team because we were not experienced in this area and we did not have any analytical devices. But we brought ourselves to start. One of the authors of the discovery I.P. Borovinskaya who was the only chemist in our laboratory but with experience in organic chemistry was responsible for the work.

For our investigation we chose two types of the process: combustion of powder mixtures of transition metals (Ti, Zr, Hf, V, Nb, Ta) with carbon (it became a research issue for a young specialist, one of the authors of our discovery V.M. Shkiro) and that of compact samples of the same powders in nitrogen (in our first experiments we found out the possibility of such processes, and it became a research direction of I.P. Borovinskaya). Unfortunately, soon V.M. Shkiro was enrolled and left our ranks. As for me, I was engaged in all the matters, but paid the main attention to macroscopic investigations. We managed to interest some other colleagues in our work: a theoretician B.I. Khaikin and his follower A.P. Aldushin, experimentalists A.K. Filonenko, Yu.M. Grigoriev. Later, a designer V.I. Ratnikov, a technologist V.K. Prokudina, specialists in materials science G.A. Vishnyakova and V.M. Maslov and others joined us.

So we started our work which was called "Self-propagating high-temperature synthesis". Sometimes I had my doubts if it was worthy dealing with the unexplored (it's the same as to go to the unknown forest in bad weather without a compass). But I got rid of them and a strong aspiration for something new prevailed. But soon I realized that these two notions – the old (combustion) and the new for us (synthesis) – could be united by our favorite word "macrokinetics" and it reassured me.

I am not going to describe this wonderful story in details – we have a lot of matters to be discussed. But its result is well-known for you because today we are celebrating the 40-th anniversary of the discovery and holding the International conference on the historical aspects of SHS.

No doubt that we thought of the question why this phenomenon had not been discovered before (it is so simple); we acquainted with some papers describing exothermic effects during heat-releasing reactions and concluded that it was impossible to develop SHS without deep understanding of combustion processes and only those specialists who were not overburdened with the attachment to specific systems and processes and possessed knowledge in the field of modern science of combustion could realize the common character of separate exothermic effects and turn their observations to harmonious ideology. It explains why SHS was developed only after the appearance of the modern combustion theory.

Almost each scientific discovery has its own portent. From this point of view, the SFP and SHS are not an exception. The authors consider that the discovery was preceded by the studies on metallothermic reactions (Beketov-Goldscmidt) and modern combustion theory (Semenov-Zeldovich) (Fig. **0.4**). **[1 – 3].**

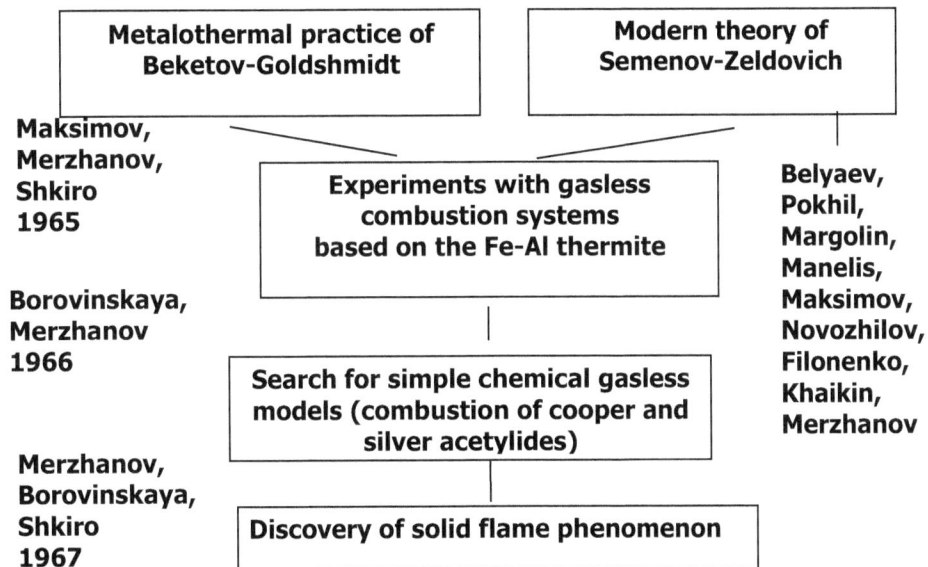

Fig. (0.4): The path of the authors to their scientific discovery.

Dedicated to My Teachers

N.N. Semenov
1896-1986

F.I. Dubovitsky
1907-1999

2

CHAPTER 1

Three Stages of SHS Development

Abstract: SHS investigation development is considered from the geographical and historical viewpoint. 3 stages are described. Within Stage 1 the work was carried out in the Department of the Institute of Chemical Physics in Chernogolovka where the scientific discovery had been made. At Stage 2 the interest to SHS arose in different cities and towns of the former USSR. Within Stage 3 SHS entered the international scene. Now SHS processes and products are being studied in more than 50 countries.

Since then SHS investigations were being developed extensively (Fig. **1.1**). At first the works were carried out only by our team in Chernogolovka. We concentrated our attention on our work and tried to understand "what's what". In 1969 at the II All-Union Symposium on Combustion and Explosion in Yerevan I reported about the solid flame phenomenon, SHS processes and our first investigation results. The report appeared to be of great interest. Our work was backed. After the symposium our efforts in the SHS development were also backed up by our immediate chief F.I. Dubovitsky.

In 1972 we lost (figuratively speaking) our sole right to SHS investigations – the first SHS hearths appeared in Yerevan, Tomsk and other towns of the former USSR. The initiators of the work were our post-graduates and trainees as well as other scientists and researchers who learned about SHS and inspired the idea of the process investigation. The State support of SHS, started in 1979 and continued up to the USSR disorganization (1992), played a great role in the development of our work.

In 1980 the SHS came out to the world arena. At first the USA and Japan, and then other countries were involved in studying these processes. The main event for the further development of this direction was an analytical review issued by J. Crider, a member of the Scientific Information Service of the USA Army, and called "Self-propagating high-temperature synthesis – a Soviet method of ceramics production" [5]. It also appeared to be a stimulating factor in the Soviet Union as the American recognition of a Soviet scientific achievement was not common those days.

So in the SHS R&D we can distinguish three stages: since 1967 – SHS in Chernogolovka; since 1972 – a Soviet period (Chernogolovka + SHS centers in other towns of the Soviet Union); since 1980 – SHS all over the world (the Soviet Union + other countries).

The work has been developed with self-acceleration and self-propagation for already 40 years. During those years a complicated and significant path has been passed from a scientific discovery to organization of industrial facilities, a lot of interesting scientific and practical results have been achieved, very many audacious organizing solutions have been accepted.

In this paper I've tried to give a generalized but thorough description of our achievements.

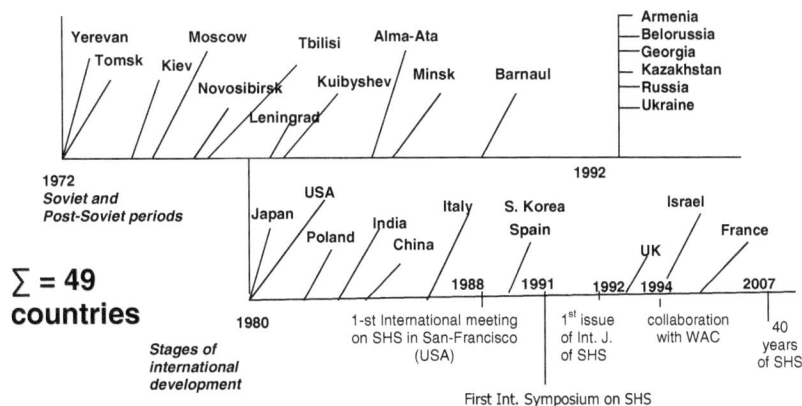

Fig. (1.1): Three stages of SHS development [4].

CHAPTER 2

Primary SHS (1967–1972)

Abstract: The results of the investigation of solid-flame processes leading to obtaining refractory compounds (metal (Group IV and VI of the Periodical Table) carbides, borides, nitrides, silicides) are described in the Chapter. The first syntheses were carried out. Simple combustion models were made. New phenomena were discovered. It led to elaboration of the scientific backgrounds of the SHS processes.

It was the name of our first stage (1967–1972). First of all we determined two tasks:

- To study SFC mechanism and regularities, to clear up the character of the solid flame phenomenon, and make simple mathematical models;

- To learn to synthesize and analyze top-quality refractory compounds (borides, carbides, nitrides, silicides).

We chose elemental systems for our investigation as they led to the formation of refractory compounds. These compounds are known to be characterized by high values of the formation energy; it means that a great amount of heat is released during their synthesis. At the same time they have high melting points. Therefore, they are ideal for realization and study of the solid flame phenomenon (Fig. **2.1**).

SHS-reaction	Adiabatic combustion temperature, K	Minimal eutectic temperature on the state diagram, K
$Ta + C \rightarrow TaC$	2734	3120
$2Ta + C \rightarrow Ta_2C$	2604	3120
$2Nb + C \rightarrow Ta_2C$	2606	2600
$V + C \rightarrow VC$	2228	2453
$Mo + B \rightarrow MoB$	2310	2220
$2Mo + B \rightarrow Mo_2B$	1788	2220
$2Mo + 5B \rightarrow Mo_2B_5$	1882	2220
$Cr + B \rightarrow CrB$	1602	1900
$W + B \rightarrow WB$	1446	2400
$2W + B \rightarrow W_2B$	1345	2400
$2W + 5B \rightarrow W_2B_5$	1580	2400

Fig. (2.1): Refractory compounds as ideal objects for solid flame combustion.

These compounds were indeed the first objects of SHS process investigation.

We coped with the assigned tasks with success. We obtained interesting results. Below you will find some of the most significant ones:

- We predicted theoretically and proved experimentally the effect of the reaction zone broadening connected with the reaction auto-retardation. It was shown that if the kinetic retardation of a reaction is not high, the process occurs according to Zeldovich (with the reaction zones being narrower than the heated-up layer). In the case of an intensive retard the reaction zone becomes wider and splits into two parts, one of which is an afterburning zone exerting no influence on the burning velocity. Such processes are typical for the solid flame phenomenon.

    The concept of broad reaction zones was further applied to other problems of combustion theory.

- A phenomenon of a spontaneously arising and self-regulating filtration of a gaseous reagent to the combustion zone in the systems "a porous gas-adsorbing reagent – an active gas at rest" was discovered. The existence of two limiting combustion modes – surface and layer-by-layer ones – was experimentally ascertained by the example of metal combustion in nitrogen. The regimes of the filtration combustion wave reflection from nonthermal obstacles were found out. They were shown to be connected with possible incomplete burning of substances in the filtration combustion wave. These results initiated a great deal of experimental and theoretical works which marked the origin of a new part of the combustion science which was called filtration combustion.

- New stationary modes of SFC wave propagation were discovered. They appeared due to the loss of stationary movement stability of a combustion plain front: autooscillating flame propagation and spin waves. These results were a stimulus for physicists, chemists, mathematicians and specialists in mechanics who were interested in nonlinear dynamics of autowave processes.

- Methods of chemical syntheses realization based on SHS processes were developed. It means that before the syntheses it is necessary to study a dependence of chemical and phase compositions of combustion products on the process and choose the terms which allow obtaining the combustion products of a required composition (optimum terms).

We mastered the methods of chemical and X-ray phase analyses (with determination of free and bound elements, impuritive oxygen content, lattice type and parameters) and it was enough for primary attestation of the products.

This approach was further used in many works.

Our initial experience proved that the SHS method required a great volume of preliminary investigation but a proper choice of optimum terms allows the syntheses to be carried out very fast and easily. It became clear that SHS is a typical science-intensive process.

Some results of our primary SHS investigations are given in Fig. **2.2–2.8**. They appeared to be scientific backgrounds which were further developed and extended.

"The Primary SHS" strengthened our wish to continue working in this direction **[4,7]**.

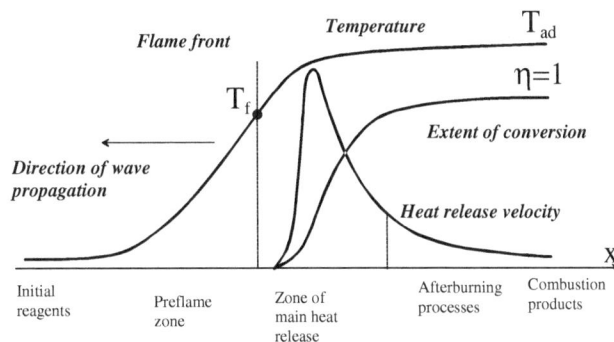

Fig. (2.2): Calculated structure of adiabatic wave (B. I. Khaikin, A.P. Aldushin, K.G. Shkadinskii, A.G. Merzhanov) [6].

Mathematical model of the process:

$$\lambda \frac{d^2 T}{dx^2} - c\rho U \frac{dT}{dx} + Q k_0\, \phi(\eta) \exp\left(-\frac{E}{RT}\right);\ U \frac{d\eta}{dx} - Q k_0\, \phi(\eta) \exp\left(-\frac{E}{RT}\right) = 0\ ;\ x = \pm\infty \qquad \frac{dT}{dx} = 0\ ;$$

$$\phi(\eta) = \exp(-m\eta) \qquad (\eta > 1) - \text{intensive kinetic breaking.}$$

T – temperature, h – conversion degree, x – linear coordinate, T_{ad} – adiabatic combustion temperature, T_f – combustion front temperature (when the heat flow from the zone of the main heat release to the preflame zone achieves its maximum), l – thermal conductivity coefficient, c – thermal capacity, r – density, Q – reaction thermal effect, E – activation energy, k_0 – preexponential factor.

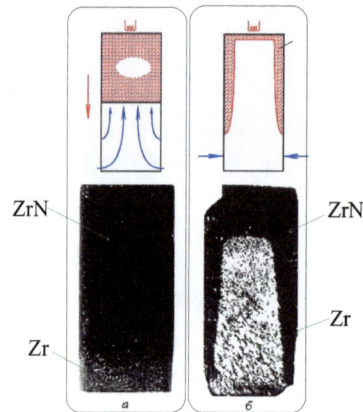

Fig. (2.3): Modes of self-controlling filtration process. a – kinetic mode, layer-by-layer combustion; b – filtration mode, layer-by-layer combustion. According to *I.P. Borovinskaya et al., 1969 – 1970.*

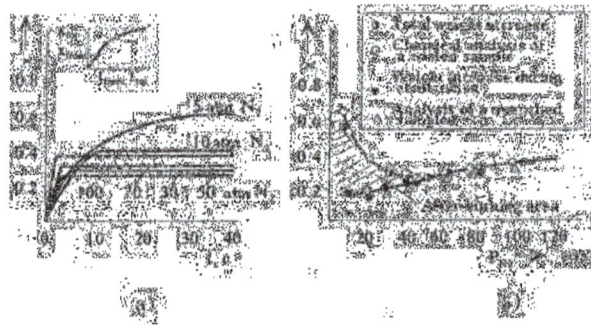

Fig. (2.4): Combustion of porous zirconium samples in nitrogen: Dependence of nitriding extent, x, on time (a – gravimetric curves) and nitrogen pressure (b) [7].

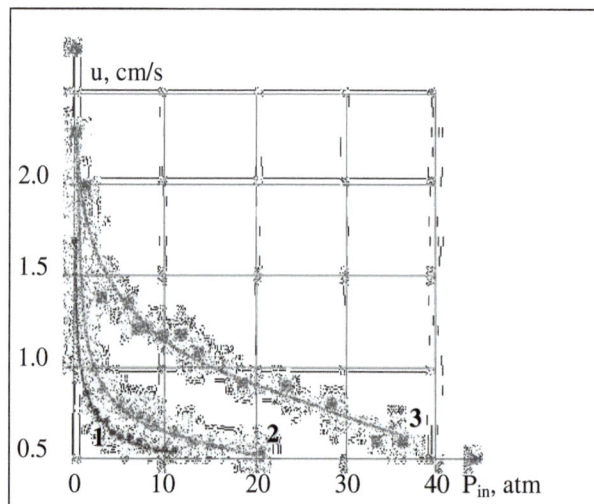

Fig. (2.5): Combustion velocity dependence on inert gas content in nitrogen for Ti-N system. Relative density ρ_{rel} = 0.41(porosity – 59%). Partial pressure of nitrogen: (nitrogen-argon medium) 1–21 atm, 2–41 atm; (nitrogen- helium) 3–41 atm.

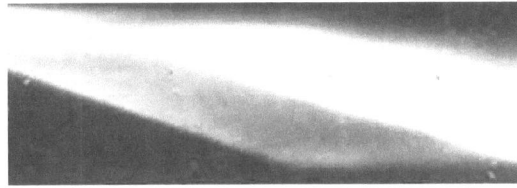

Fig. (2.6): Photographic record of hafnium combustion in nitrogen: successive propagation of two combustion fronts.

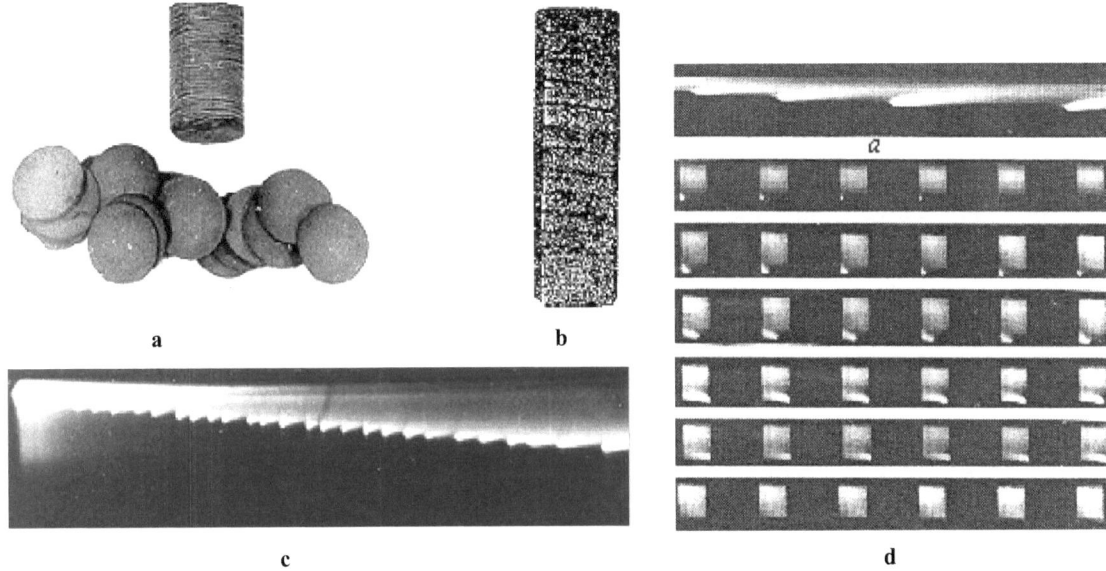

Fig. (2.7): New modes of combustion wave propagation (I.P. Borovinskaya, A. K. Filonenko, A.G. Merzhanov, 1972). (a) Lamellar product of oscillating combustion, (b) helical trajectory of hot spot on the side surface of combustion product, (c) rapid movie picture of oscillating combustion, and (d) still frames of spinning combustion (Hf in nitrogen) [8].

	Metal content, wt %		Nonmetal content, wt %		Crystal structure	H_μ, $kg \cdot mm^{-2}$
	bound	free	bound	free		
TiC	80	$< 10^{-3}$	19.8	$9 \cdot 10^{-2}$	cubic	2900
ZrC	88.2	$< 10^{-3}$	11.4	$< 10^{-2}$	–"–	3040
NbN	86.9	$< 10^{-3}$	13.0	–	–"–	1670
TiN	77.6	$< 10^{-3}$	21.4	–	–"–	1800
TaN	91.8	–	7.5	–	–"–	3200
BN	43.8	0.2	55.7	–	hexago-nal	–
TiB$_2$	68.6	–	31.0	10^{-1}	–"–	3500
HfB$_2$	89.0	–	10.7	$< 10^{-2}$	–"–	2890

Fig. (2.8): The first refractory compounds produced by SHS (I.P. Borovinskaya, A.G. Merzhanov, 1971) [2].

CHAPTER 3

Favorable 70-S: Investigation Branching

Abstract: Active development of the investigation started in the 1970-s. The results obtained within Stage 1 were under study, some new directions appeared. Mechanisms of SHS product formation were realized.

After summing up at the initial stage of our investigations, we realized that we could advance in SHS. And we did it. We had a good team in Chernogolovka. Some of our colleagues who had been engaged in other research themes (as for us, we dealt with different directions of macroscopic kinetics) turned their attention to SHS, new members joined us. We worked with great pleasure, with the common wish to perceive the new process, we worked in harmony without keeping our fresh ideas from the others, without reckoning with time; we argued a lot. And as a result SHS clued allowing us to evolve itself.

Almost each question touched upon at the initial stage found not one but several solutions and we began speaking about the investigation branching. Fig. **3.1–3.3** are demonstrating it. It is obvious that a set of problems under study became rather wide. Some of them followed the available methodological and ideological schemes (but with other objects), the others gave new opportunities to the researchers. The first group included formalized and semi-empirical investigations in chemical syntheses of new objects, the second one dealt with thorough studying of chemical reaction mechanisms with revelation of intermediate stages and separation of intermediate products.

Methods of experimental diagnostics	<ul><li>video recording</li><li>thermometry</li><li>calorimetry</li><li>electrothermography</li></ul>
Experimental investigations: from regularity of front propagation to study of:	<ul><li>wave structure</li><li>mechanisms of infiltration and layer-by-layer combustion</li><li>kinetics of high-temperature interaction between metals and gases</li><li>new combustion processes (combustion of metals in hydrogen)</li></ul>
Theory and mathematical modeling: from the simplest 1-D models to:	<ul><li>fundamentals of infiltration combustion</li><li>conception of stability loss and relationship with phenomena of non-equilibrium combustion</li><li>development of combustion modes (complex and more appropriate to experimental conditions)</li><li>conception of equilibrium and non-equilibrium mechanisms of SHS</li></ul>

Fig. (3.1): Fruitful 70-s: Macrokinetics.

Reactants and raw materials: from elemental mixtures to:	<ul><li>Mixtures containing magnesium as a reducing agent</li><li>Ferroalloys and master alloys</li><li>Various compounds as reactants and adjust additions</li></ul>
Synthesis: from refractory compounds to:	<ul><li>Thermal unstable compounds</li><li>Nitride ceramics</li><li>Tungsten –free alloys</li><li>Nitrided ferroalloys</li><li>Various compounds</li></ul>
Development work: from syntheses in a laboratory to pilot-scale production	<ul><li>Powder technology</li><li>Gas-statting technology for production of ceramic items with high and low porosity</li><li>Forced compaction of hard alloys</li><li>Technology of high-temperature melts or SHS metallurgy</li></ul>

Fig. 3.2: Fruitful 70-s: Chemistry for SHS processes and products.

Analysis: from primitive attestation to:	• X-ray diffraction analysis • neutron diffraction analysis • physical-chemical analysis • metallography
Application of SHS products: from laboratory syntheses to practical application	• Production of high-temperature heaters • Production of abrasive materials (pastes and abrasive powders based on SHS-TiC) • Application of nitrided ferro-vanadium for alloying steel • Production of ceramic insulators (sleeves) in furnaces of oriented crystallization
New SHS centers in the former USSR: from Chernogolovka to	• Yerevan • Tomsk • Kiev • Moscow • Leningrad • Ulugbek

Fig. (3.3): Fruitful 70-s: Analysis, application, new participants.

Let's consider some investigation results of the scientists from Chernogolovka.

Fig. **3.4** shows some generalized data of realization of SHS processes and their characteristics. It is obvious that SHS can be referred to extreme chemical processes.

Metal particle size	$5 - 150$ μm
Nonmetal and compound particle size	$0.1 - 1.0$ μm
Sample diameter	$5 - 30$ μm
Sample length	$2 - 5$ diameters
Sample porosity	$70 - 40$ %
Initial temperature	$293 - 800$ K
Gas pressure	From vacuum to 150 bar
Front propagation velocity	$0.1 - 20$ cm/s
Maximum combustion temperature	$2300 - 3800$ K
Heating velocity in the wave	$10^3 - 10^6$ K/s
Ignition capacity	$10 - 200$ cal/cm^2·c
Ignition delay	$0.2 - 1.2$ s
Ignition temperature	$1200 - 1600$ K

Fig. (3.4): Typical parameters of green mixtures, experimental terms and SHS-process.

Specialists in combustion know that a glowing front observed by means of optical methods is a front edge of the wave followed by the combustion zone in which very important processes take place; they determine the wave movement character and velocity. But what about this zone structure? Some information can be got by calculation methods. But is the calculation information adequate to actual data? New methods of experimental diagnostics allowed us to measure a temperature profile in the combustion wave. It supplied a lot of useful information for studying the combustion mechanism.

Figs. **3.5, 3.6, 3.7** present the information showing how to organize an experiment in order to achieve the correspondence of experimental and calculation data by the temperature profile which is the most important feature of SHS processes.

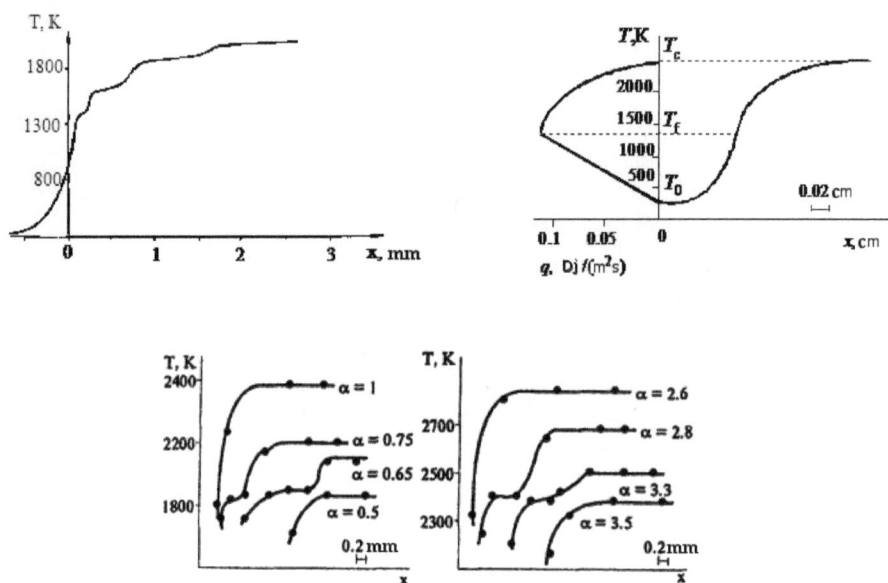

Fig. (3.5): A temperature profile measured by thermocouples at combustion of 5Ti + 3Si (A.A. Zenin) (1), processing of a simple temperature profile by Zenin's method: calculation of heat flows in Nb+2B combustion wave (2), temperature profiles measured by optic-spectrum method (V.M. Maltsev, *et al.*) (3) [9].

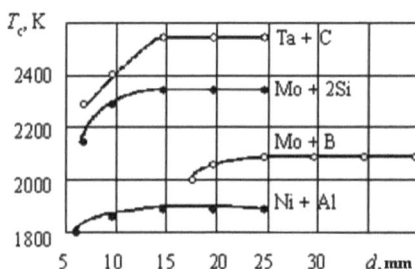

Fig. (3.6): Thermodynamics, calorimetry, and combustion temperature.

The diagram shows the dependences of the maximum combustion temperature on cylindrical samples' diameter. In the case of rather big diameters the temperature does not depend on them. It is an adiabatic mode in which heat losses into the surroundings can be neglected. When the diameter is lower than a specific value, the combustion velocity decreases due to the combustion temperature decrease resulted from heat losses. The reaction heat can be measured by SHS calorimeter designed by L.N. Galperin and L.B. Mashkinov.

	Product	Method of combustion temperature measurement, K			Heat of formation, kJ/mol		Calculated adiabatic combustion temperature, K
		Thermo-couple	Micro-thermo-couple	Pyro-meter	SHS-calorimeter	Literature data	
Ti + B	TiB	2500	2570	2500	167.2± 3.3	158.8	2500 ± 50
Ti + 2B	TiB_2	3190	-	3150	298.9± 5.8	293.0 – 293.5	-
Ta +2B	TaB_2	2300	2670	-	198.5± 4.2	193.5 – 217.4	2280 ± 50

Fig. (3.7): Thermodynamics, calorimetry, and combustion temperature. Experimental and calculated combustion parameters for some SHS-systems.

Direct synthesis from elements (I.P. Borovinskaya, V.M. Shkiro)

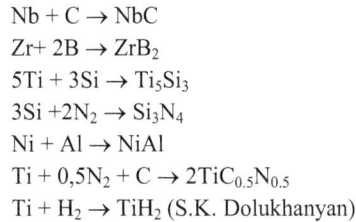

$$Nb + C \rightarrow NbC$$
$$Zr + 2B \rightarrow ZrB_2$$
$$5Ti + 3Si \rightarrow Ti_5Si_3$$
$$3Si + 2N_2 \rightarrow Si_3N_4$$
$$Ni + Al \rightarrow NiAl$$
$$Ti + 0,5N_2 + C \rightarrow 2TiC_{0.5}N_{0.5}$$
$$Ti + H_2 \rightarrow TiH_2 \text{ (S.K. Dolukhanyan)}$$

Direct synthesis from compounds (V.V. Boldyrev)

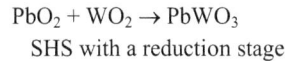

$$PbO_2 + WO_2 \rightarrow PbWO_3$$

SHS with a reduction stage
(I.P. Borovinskaya, S.S. Mamyan, V.I. Yukhvid)

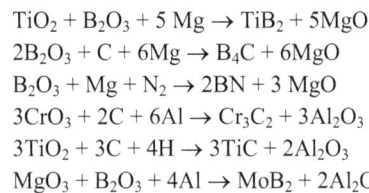

$$TiO_2 + B_2O_3 + 5\,Mg \rightarrow TiB_2 + 5MgO$$
$$2B_2O_3 + C + 6Mg \rightarrow B_4C + 6MgO$$
$$B_2O_3 + Mg + N_2 \rightarrow 2BN + 3\,MgO$$
$$3CrO_3 + 2C + 6Al \rightarrow Cr_3C_2 + 3Al_2O_3$$
$$3TiO_2 + 3C + 4H \rightarrow 3TiC + 2Al_2O_3$$
$$MgO_3 + B_2O_3 + 4Al \rightarrow MoB_2 + 2Al_2O_3$$

Chemical mechanism for tantalum combustion in nitrogen
(I.P. Borovinskaya, A.N. Pitulin, V.Sh. Shekhtman):

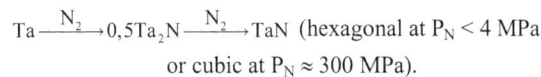

$$Ta \xrightarrow{\;N_2\;} 0,5Ta_2N \xrightarrow{\;N_2\;} TaN \text{ (hexagonal at } P_N < 4 \text{ MPa}$$
$$\text{or cubic at } P_N \approx 300 \text{ MPa).}$$

Fig. (3.8): Chemical SHS-reactions.

During the initial stage of SHS the processes in elemental systems were studied, it was explained by the terms imposed by the solid flame phenomenon. Then the group of reactions realized (Fig. **3.8**) at the SHS mode became wider, we began studying the processes involving some compounds as initial reagents. Among them were metallothermic reactions which were rather popular in metallurgy those years but their opportunities became significantly wider with application of new devices (vessels under pressure, centrifugal devices).

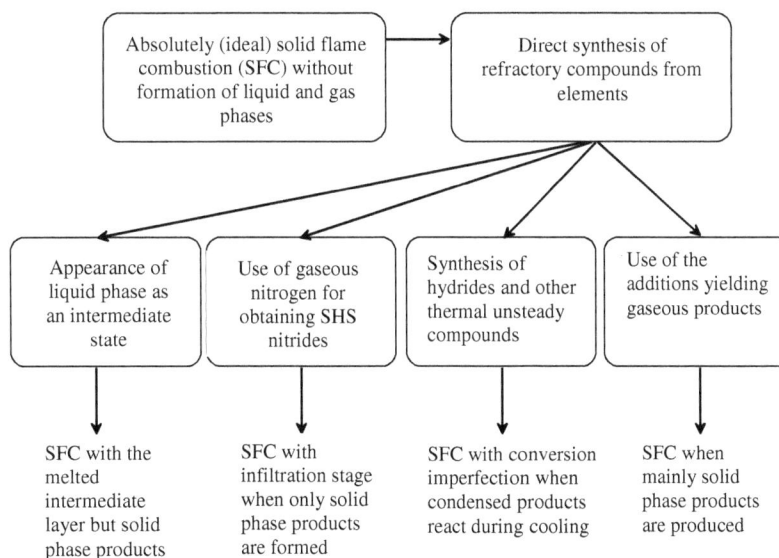

Fig. (3.9): "Solid flame" combustion (SFC) and SHS. SFC is a type of combustion with completely or partially solid products.

It became clear that the notion of the scientific discovery and our aim to connect SHS with the phenomenon of solid flame restricted the possibility of SHS as a synthesis method. At the primary stage we could maintain the links between SFC and SHS (Fig. **3.9**) but then their roads dispersed, SHS tore itself away from the solid flame and began its independent life.

A lot was done to realize the character of such phenomena as combustion front autooscillations and spin waves. We had considered them as independent phenomena but then realized that they were closely connected with the character of stationary flat wave instability and appeared after the loss of stability. These processes were called unsteady-state combustion. But the better name for them is combustion modes in the zone of flat wave instability.

I'd like to give you one more example. It concerns two different mechanisms of a substance conversion during combustion processes (Fig. **3.10**):

a)   Equilibrium mechanism assumes that SHS reaction proceeds in the mode of reactive diffusion within the zone of the main heat release in the vicinity of combustion front. It was developed using the conceptions of the interaction mechanisms under high temperatures.

b)   Non-equilibrium mechanism was developed due to analysis of experimental results of zirconium and titanium combustion in nitrogen at high pressures; the combustion reaction yields not a final but a non-equilibrium intermediate product which due to the subsequent dissociation gives the final product.

The experimental diagnostics of these combustion modes became possible after development of time-resolved XRD analysis [8 – 11].

Equilibrium Mechanism *Conception of Khaikin-Aldushin-Merzhanov*	*Non-equilibrium Mechanism* *Conception of Borovinskaya*
The final product is formed during the combustion reaction (a typical example: the combustion reaction occurs under the mode of reaction diffusion)	The final product is formed far behind the combustion front as a result of non-equilibrium state disintegration of the combustion product

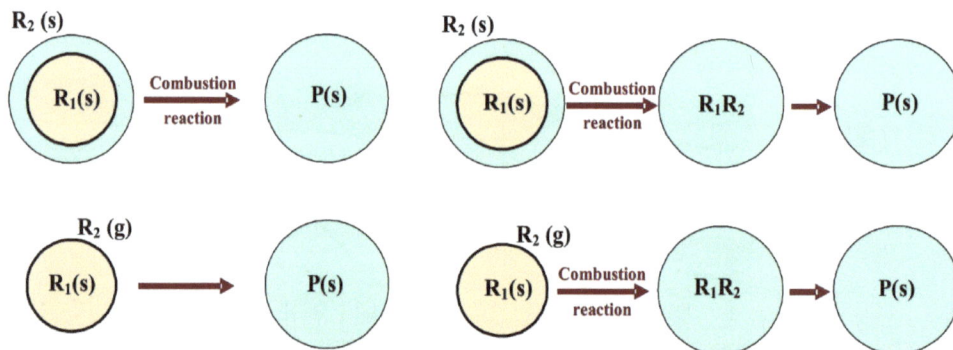

Fig. (3.10): Equilibrium (a) and non-equilibrium (b) mechanism of SHS reactions.

CHAPTER 4

Favorable 70-S: Geographical Branching

Abstract: The Chapter describes the organization of the SHS work in the Ukraine, Armenia, Tomsk, etc. The first results were obtained in the industrial assimilation of SHS. Scientific-and-industrial cooperation between different institutions producing abrasive pastes was established. The fruitful work in the 1970-s proved that SHS can be a background for development of new scientific and technological directions.

One of the most important results was "geographical" branching of SHS, i.e. self-propagation of SHS throughout the former USSR. In the 70-s the Scientific Center of the Academy of Sciences in Chernogolovka ceased to be the only place of SHS investigation. New research centers were founded in Yerevan (the Laboratory of chemical physics at the Armenian Academy of Sciences), the Research Institute of Applied Mathematics and Mechanics at Tomsk State University, the Institute of Materials Science in Kiev, the Institute of Nuclear Physics of the Uzbek Academy of Sciences in Ulugbeck, etc. In each case the beginning was different but everywhere the researchers from Chernogolovka were very active and promoted the work outside our town.

4.1 SHS IN ARMENIA

The first republic which was involved in SHS formation was Armenia. The idea to develop SHS belonged to my friend Prof. L.O. Atovmyan, an outstanding scientist in crystal chemistry, the Head of the Laboratory of the Branch of the Institute of Chemical Physics. After realizing that SHS was a rich area of R&D, he said to me: "It should be done in Armenia". The same words were told to the prominent scientist – academician N.S. Yenikolopov who had a conversation about it with B.A. Muradyan, the Chairman of the Council of Ministers of the Armenian SSR. Being a chemist, B.A.Muradyan appreciated the significance of SHS at once and organized my report at the Meeting of the Council of Ministers. It resulted in issuing the Regulation which entrusted academician A.B. Nalbandyan, Director of the Laboratory of Chemical Physics, with organization of SHS R&D. The work was financially supported, and from the very beginning it was being carried out in close contact with the researchers from our laboratory in Chernogolovka (Fig. **4.1.1**).

L.O. Atovmyan   **N.S. Enikolopov**   **A.B. Nalbandyan**

S.K. Dolukhanyan   **S.L. Kharatyan**

Contribution of Chernogolovka

In Chernogolovka	In Yerevan
Training of young researchers	Definition of a problem and discussion of results
(Yu.M. Grigor'ev)	(A.G. Merzhanov, I.P. Borovinskaya)

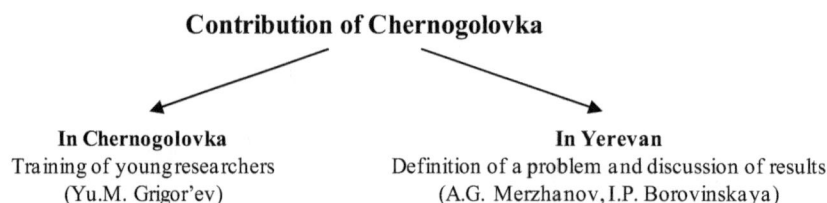

Fig. (4.1.1): SHS in Armenia.

I became the informal Head of the work. One of my active assistants was the SHS leader in Chernogolovka – I.P. Borovinskaya. There was a great difficulty – Armenia was not ready to start working in SHS because there were no specialists in this field (it is clear), moreover, our colleagues there did not carry out any research close to ours – combustion of condensed systems. We decided to eliminate the lack of knowledge and experience in two ways. Without delay we started assimilating SHS in their laboratory and at the same time their young specialists were sent to Chernogolovka to study the process. The result was excellent. Among the people involved in the work there appeared a leader – Seda Dolukhanyan, a clever, purposeful, industrious woman. She collaborated with Inna Borovinskaya who came to Yerevan rather often. Inna passed our work in silicide synthesis (started in Chernogolovka) to our Yerevan colleagues, and we hoped it would become the main direction of their activity. When organizing SHS research centers in Yerevan and in other places, we always aimed at imparting their own features to them. But silicides did not become the characteristic feature of Armenian SHS. The fate favored us with another solution of the problem. At first we did not have any pure reagents and used only available ones. But they did not satisfy our ambitions. Inna Borovinskaya suggested introducing hydrides into green mixtures in order to obtain purer products. But it was difficult to get hydrides too. We decided to synthesize hydrides ourselves by heating metal powders in hydrogen medium. The result was unexpected. When heated, the powders flared and turned into hydrides. It was enough for us to understand that not only refractory compounds but thermally instable hydrides could be obtained by SHS. Thus, the investigation (and then hydride application) allowed our Armenian colleagues to find their own niche in the SHS problem.

Training in Chernogolovka was a success too. The young specialists joined my department of engineering chemical physics which had been founded with the Polytechnic Institute in Kuybyshev. The specialists took a course of lectures on macroscopic kinetics (including SHS). The most active was Suren Kharatyan. He became a post-graduate of Prof. Yu.M. Grigoriev and defended his thesis on kinetics of high-temperature interaction of transition metals with nitrogen.

When the Armenian specialists returned to Yerevan, A.B. Nalbandyan organized the SHS department in his laboratory. It consisted of two sections: one of them headed by S.I. Dolukhanyan was engaged in hydride synthesis, chemistry and technology, the other run by S.L. Kharatyan – in macroscopic kinetics. Our relations with Armenia were very close. The successful development of SHS R&D was noticed and the laboratory headed by Nalbandyan was reorganized to an Institute.

Let's note two results achieved in the 70-s (Figs. **4.1.2, 4.1.3**). In addition to synthesis of ordinary hydrides, some works in obtaining intermetallic hydrides were carried out. The possibility of such synthesis depends on combustion temperatures. If the temperature is high, an intermetallic compound decomposes to the constituents. At lower temperatures it does not decompose and forms an intermetallic hydride.

Fig. **4.1.3** answers the question which has been of great interest for us from the very beginning: How and why can hydrides be formed by SHS? Hydrides are unstable compounds, and when heated, they easily dissociate (not at very high temperatures but at the values which are lower than the calculated adiabatic ones). It is obvious that when comparing the hydride dissociation curve with the combustion adiabatic line, we get the information what products are formed in the combustion wave. In the point of their intersection we obtain the data on the combustion temperature and the hydration degree. For most of the cases concerning hydrides the intersection point belongs to the high-temperature branch of the dissociation curve. It means that hydride is not formed in the combustion wave. If we carried out annealing, we would synthesize hydrogen solid solution in a metal. Hydride is formed at the cooling process. A temperature decrease shifts the equilibrium in the metal–hydrogen system toward hydride and the combustion product is additionally enriched in hydrogen: at first a nonstoichiometric hydride is formed and then its composition approaches a stoichiometric value.

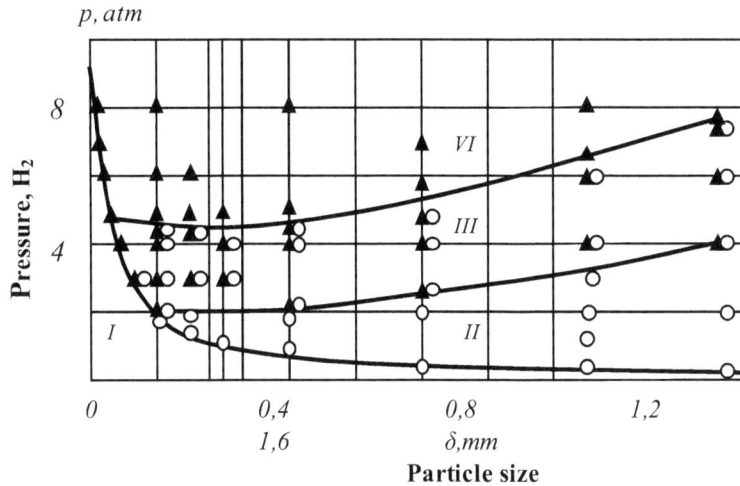

Fig. (4.1.2): Zr_2Co combustion in hydrogen (S.K. Dolukhanyan, A.G. Akopyan, A.G. Merzhanov, 1981).

I   – no combustion due to significant heat losses;

II – intermetallic hydride is synthesized by the scheme: $Zr_2Co + H_2 \rightarrow Zr_2CoH_2$ (combustion temperature < 550 °C);

III – partial decomposition of the intermetallic compound at hydrogenation: $Zr_2Co + H_2 \rightarrow Zr_2CoH_2 + Zr_2Co + ZrH_2$ (combustion temperature >550°C);

IV – complete hydrogenolysis: $Zr_2Co + H_2 \rightarrow ZrH_2 + Co$.

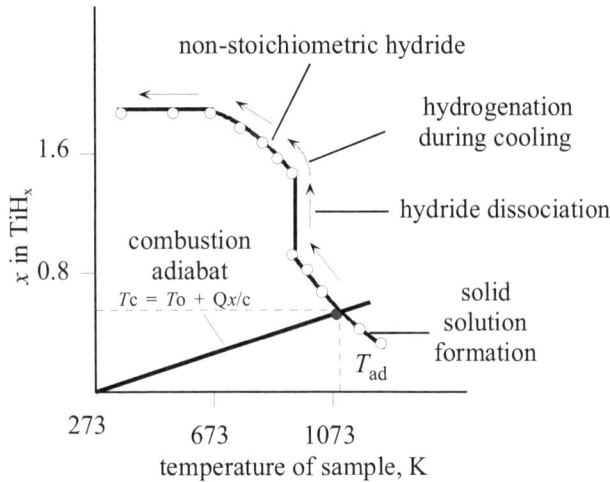

Fig. (4.1.3): Formation of thermally unstable hydrides at SHS (S.L. Kharatyn, A.G. Merzhanov). At the maximum combustion temperature an intermediate product, it is a solid solution, is formed. At cooling the product is saturated with hydrogen and turns to hydride.

The first work in synthesis of silicides carried out by our Armenian colleagues with us was also successful (Fig. **4.1.4**). The technology of molybdenum disilicide powder was introduced in Kirovakan plant, demonstrated a high output and was awarded to the Armenian State Prize. Thus, the foundation was laid and it allowed the Armenian branch of SHS to withstand the test of time and hardship and to keep the activity of the scientists.

Efficiency:

- an improved quality;

- released energy of induction furnaces;

- a decrease in the number of the staff and operation area;

- an increase in the production output;

- a decrease in the production cost [10,15].

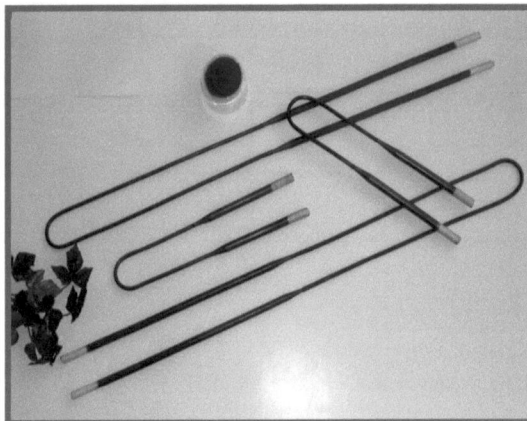

Fig. (4.1.4): High-temperature heating elements based on SHS $MoSi_2$ and made at Kirovakan plant (Armenia).

4.2. SHS in Tomsk: the 70-s

The work in Tomsk where the third SHS center was founded was organized in a different way.

Our post-graduate student Yury Maksimov, who was preparing his thesis but not in SHS, was very interested in our work in Chernogolovka. He liked the new process and after defending his thesis, he decided to go back to his native town – Tomsk, where he had graduated from the University, and to develop SHS there. Tomsk is famous for the realized scientific investigations and for the well-known combustion school of Prof. Vilyunov. As to specialists, the situation in Tomsk was much better than in Yerevan but there were no financial means assigned for SHS. But their aspiration played the main role.

After coming back to Tomsk, Yu.M. Maksimov organized a laboratory and began working in the Research Institute of Applied Mathematics and Mechanics of Tomsk State University (Fig. **4.2.1**). He got in touch with V.I. Itin, who worked in Siberian Physical & Engineering Institute and had some experience in studying exothermic reactions, invited a young but experienced researcher of experimental diagnostics of combustion processes Yu. S. Nayborodenko, gathered young specialists and students, and strengthened the relations with the colleagues from Chernogolovka.

We realized that his work would be recognized in Tomsk if he made something prominent. He appeared to be lucky. I.P. Borovinskaya decided to pass him her work in nitriding ferrovanadium; V.N. Lebedev, a representative of one of the Ministries suggested developing the SHS technology of nitrided ferrovanadium and assimilating it in the plant "Izhstal". There was a deficit of cold-resistant steels in the country, and nitrided ferrovanadium was a good addition to steel which provided such a property.

The result was marvelous. Due to our first investigations in Chernogolovka we knew that ferrovanadium could be obtained by SHS. When burning it up in nitrogen, Yu.M. Maksimov got an excellent product: during the SHS process the entire vanadium as a part of the ferroalloy was nitrided and converted to nitride (Fig. **4.2.2**).

Moreover, nitrogen from the SHS product appeared to be completely consumed by steel. Therefore, SHS allowed us to obtain a perfect nitrided ferrovanadium. Besides, it was well known that the SHS technology was simple but highly productive and consumed little power.

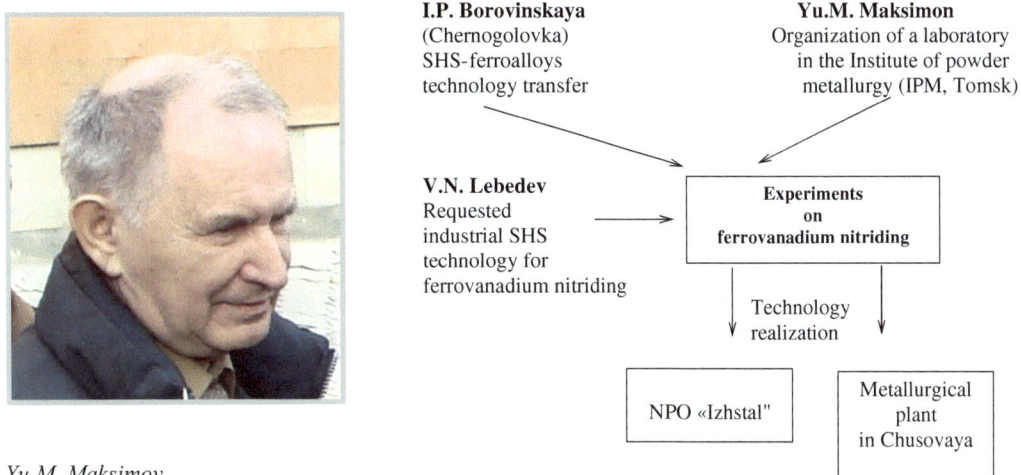

Yu.M. Maksimov

Fig. (4.2.1)": SHS in Tomsk.

Nitrided Ferrovanadium

Parameters	SHS	Produced in Furnaces at the Ferroalloys Plant in Zaporozh'e	USA, "Union Carbide"
Basic service parameters			
Nitrogen content, mass %	8–12	2–11	6–12
Vanadium content, mass %	42–50	34	78–82
Production			
Power consumption, kW · h for 1 ton production	0,5	1100	
Efficiency, kg/h	200	30	
Technological cycle time, h	1,5–2,0	70	
Temperature of nitriding, °C	Up to 1600	Up to 1200	1200
Application in steel alloying			
Nitrogen consumed by steel, %	95–100	50–75	60–80
Vanadium consumed by steel, %	85–90	50–80	60–80
Reduction of alloy consumption, times	2–3		

Fig. (4.2.2): Nitrided ferrovanadium produced in Tomsk.

It was a great success. Then the SHS technology of this ferroalloy was also assimilated in the metallurgical plant in the town of Chusovaya.

The colleagues both in Chernogolovka and Tomsk realized that it was their common victory and it encouraged them in their wish to cooperate.

Nitriding of ferroalloys became one of the main research directions of Yu.M. Maksimov's team [16]. They studied the nitriding processes of other ferroalloys. They managed to solve another complicated task, i.e.

the production of nitrided ferrochromium. Yu.M. Maksimov appeared to be a lucky man. The researcher responsible for the work was Mansur Ziatdinov, a thoughtful, industrious and hard-working man.

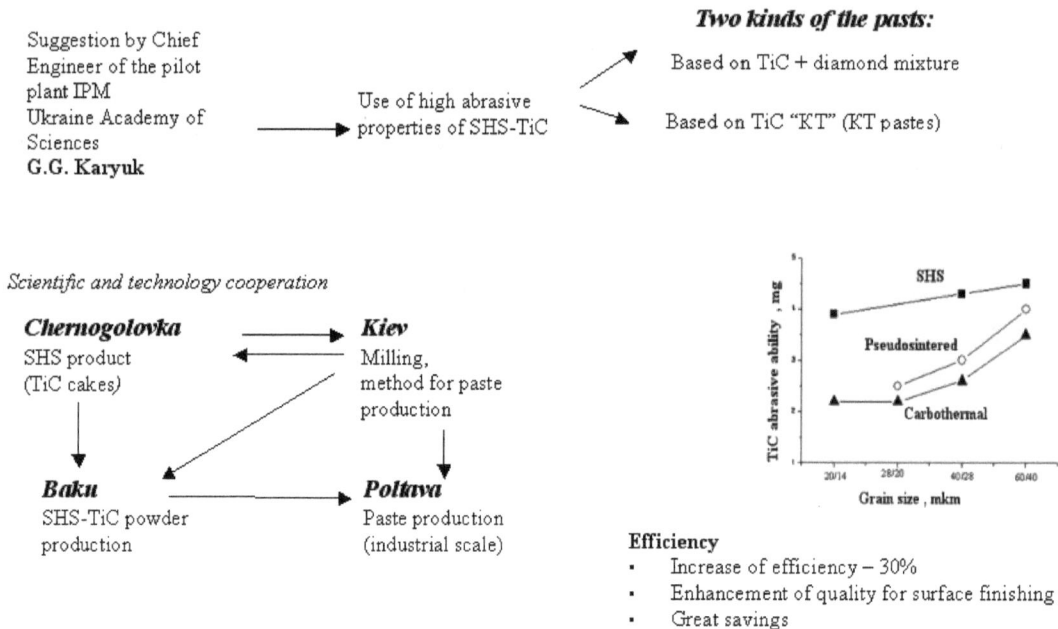

Fig. (4.3.1): Cooperation with the Ukrainian partners in production of abrasive pastes based on SHS-TiC.

Yury Maksimov's team was small but very creative. They were anxious for something new. From this very standpoint they carried out various investigations. And the results did not keep them waiting. It was the laboratory of Yury Maksimov where spin waves in the mixtures of "solid–solid" were observed for the first time (before, they had been noticed only in "solid–gas" systems).Besides, it was understanding of these experimental data which allowed them to develop a proper conception of the thermal instability of combustion in SHS processes [18,21].

In some products synthesized in the magnetic field and possessing magnetic properties they discovered oriented domain structures providing a very high coefficient of thermal conductivity along the wave movement. They noticed that polymorphous transformations occurring in the combustion wave of the system "solid-gas" caused not only a decrease but also an increase in the combustion front velocity. It was a great pleasure for us to have such a "young brother".

4.3. Our Cooperation with Ukraine

It started when G.G.Karyuk, the Chief Engineer of the engineering and design department of the Institute of Materials Science of the Ukrainian Academy of Sciences, came to Chernogolovka. He said the following: "A small piece of your SHS titanium carbide fell into my hands. I measured its abrasive ability and it appeared to be higher than that of our carbothermal product. I am engaged in development of abrasive pastes based on titanium carbide. They allow substituting partially or completely very expensive diamond in the pastes used for polishing surfaces of ferrous and non-ferrous metals. Let's solve this problem together. You will supply titanium carbide cakes; we shall mill them and make pastes".

I liked such an approach. I felt flattered since the Ukrainian materials science was famous in the former USSR and we were only beginners. I was about to agree but asked shyly: "How much titanium carbide would you like?" The answer was: "Not much. About 10 tons". Silence. I don't want to describe all the emotions following this dialogue but frankly speaking I agreed only when Inna said: "If it is necessary, I am ready to make it". For our new pilot installation it was a good test but it coped with the task very well.

G.G.Karyuk appeared to be an excellent specialist with a perfect comprehension of engineering and materials science. A man of his word. It was pleasant and interesting to deal with him. His colleagues manufactured a lot of pastes from our carbide, sent them to different plants for testing, received the replies, studied them and after all determined the technical and economical efficiency. It was great. We concluded that it was necessary to make a production line. And without any State decisions Karyuk and I organized a scientific production association (Fig. **4.3.1**).

G.G.Karyuk was a real expert in his field and I learned from him a lot. It seemed to me that we had done too little in our joint work though the fundamental contribution to that cooperation belonged to us. That is why I was very excited when I succeeded to explain why our pastes were so efficient, why they could be used for both operations – grinding and polishing – simultaneously. The main reason was the product structure. It consists of fine crystallites closely connected with each other and pores. When the cake is ground, agglomerate particles form, which are used as the paste filler. First, these particles grind the surface then they are destroyed forming crystallites – finer single-crystalline particles which carry out polishing. Thus, both processes – grinding and polishing – are realized during one operation. It explains the high efficiency [22].

We continued our cooperation but unfortunately, there was not another significant task to be solved by SHS.

Nevertheless, the Ukrainian SHS was further developed. I mean titanium compound synthesis and organization of its pilot production by V.A. Drozdenko and P.M. Prozorov, synthesis of silicon carbide of high hardness by G.G.Gnesin, and SHS of some chalcogenides.

4.4. SHS IN OTHER ORGANIZATIONS OF THE FORMER USSR

In the 70-s some other organizations became interested in SHS. All the works were carried out with our team in Chernogolovka irrespective of the fact who had suggested the idea.

I'd like to dwell on two partnerships. One of them is connected with neutron diffraction analysis of SHS products (especially, non-stoichiometric carbides, nitrides and hydrides). A respectable center of neutron diffraction analysis was the Institute of Nuclear Physics of the Uzbek Academy of Sciences. By that time our products had not been thoroughly analyzed yet. We did not go beyond minimum attestation of our products. That is why finer structural investigations were of interest for us. Neutron diffraction analysis allowed studying a fine structure of a crystal lattice and thus, realizing structural peculiarities of SHS products. Non-stoichiometric compounds – transition metal carbides, nitrides, and hydrides were of particular interest since it was clear a priori that they had a lot of possibilities to arrange nonmetal atoms and vacancies. Two questions were especially significant for us:

1. What is the degree of ordering in non-metal sublattice of our products and can they form superstructures, i.e. achieve the highest degree of ordering?

2. What is the imperfection in metal sublattice?

These questions were solved. The brightest result was obtained in our joint work in neutron diffraction analysis of superstoichiometric tantalum nitride (Fig. **4.4.1**). The main conclusion was: superstoichiometricity of the product is explained by its defective sublattice and nitrogen atoms take octahedral positions and form the superstructure [**23, 24**].

Thus, the formula of the synthesized compound must not be written as $TaN_{1.2}$, it should be presented as $Ta_{0.823}N$ or Ta_5N_6.

Irkin Karimov and his colleagues studied some other SHS products. Unfortunately, we did not understand the relationship between the structure and the synthesis mode. Now it is clear that we should study the dependence of a degree of ordering on thermal relaxation (cooling) time of a burnt-up sample. If the time is long an equilibrium product is formed; if it is short the product can be nonequilibrium with various degrees of ordering. Of course, it would be wholesome to continue that work with our present-day experience but it's very difficult to organize such works now.

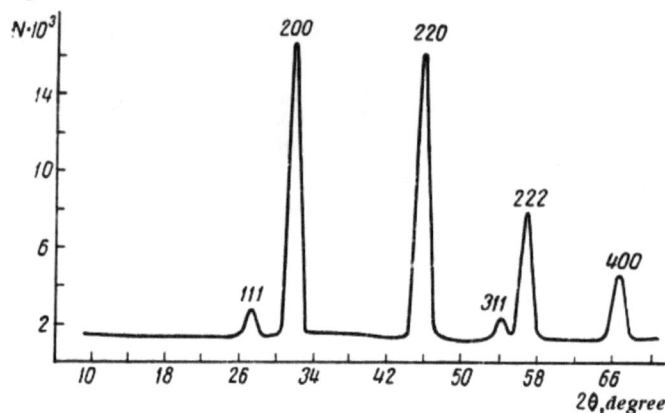

Fig. (4.4.1): Neutron-diffraction pattern of tantalum nitride. Cubic tantalum nitride was obtained by tantalum combustion in gaseous nitrogen at P>300 MPa in a hermetically sealed vessel. It was characterized by superstoichiometric composition $TaN_{1.2}$. Neutron-diffraction analysis proved than the superstoichiometry was caused by the incomplete sublattice of tantalum atoms, so the formula can be written as follows $Ta_{0.83}N$ or Ta_5N_6. The conclusion was proved by the density measurements: the calculated X-ray density – 15.95 g/cm^3, picnometric density – 13.26 g/cm^3.

Another example describes our joint efforts with A.S. Shteinberg who was a Laboratory Head at the State Institute of Applied Chemistry in Leningrad (St. Petersburg now). By that time we'd already collaborated with A.S. Shteinberg in macroscopic kinetics (thermal explosion, ignition, linear pyrolysis, geyser processes, etc.). It was natural for us to start a new stage of our cooperation in SHS. One of his works carried out at that time was investigation of nondetachable joining of refractory metals. It was an important task which could not be solved by conventional approaches. The result was a new method of welding which was based on the phenomenon of electrothermal explosion studied by A.S. Shteinberg previously. It was called SHS welding. It was not developed thoroughly (in comparison with other technological types of SHS) but allowed solving some engineering problems in Russia and abroad.

Results

Many scientists from various towns and republics of the former USSR became interested in our new direction. Due to our joint efforts some methods of experimental diagnostics, combustion theory and mathematical modeling were further developed. New chemical compounds were synthesized (not only refractory materials but also non-stable ones; in addition to elemental mixtures some compounds were used as starting reagents).

Some works on kinetics of high-temperature interaction in material-forming chemical processes were started.

Investigations of SHS product structure and properties got under way. The technological approaches of SHS powder production and direct obtaining of materials and items were tested. The first steps in practical application of SHS processes and products were taken.

The main result: the scientists engaged in SHS concluded that it was the process which allowed them to advance and contribute to the technological progress **[12 – 16]**.

CHAPTER 5

Government Support of SHS (1979 – 1992)

Abstract: In 1979 the State support of SHS started. Different statesmen visited our town and got acquainted with the method. Also the author pays great attention to the foundation of the inter- branch scientific-and-technical unit which was called Termosyntez. It united scientific, designing and manufacturing organizations. During its work the Unit demonstrated the ability of SHS to solve some important technological problems. But after the USSR disintegration the Unit collapsed. The Chapter describes the results obtained during the existence of the Unit.

5.1. N.K. BAYBAKOV, CHAIRMAN OF USSR PLANNING DEPARTMENT: A VISIT TO CHERNOGOLOVKA

In 1979 we had a real surprise. Chairman of the USSR Planning Department N.K. Baybakov came to Chernogolovka. He wanted to see the achievements of the Branch of the Institute of Chemical Physics, USSR AS. We told him about our new technology of tungsten-free hard alloys; the process was based on forced compaction (forced densification) of hot combustion products (**5.1.1** and **5.1.2**). He liked our work and suggested preparing and issuing the Resolution of Council of Ministers on further development of SHS. Certainly, we agreed. Few days later one of his assistants and I drew up a draft resolution and submitted it to the Council of Ministers for consideration. The draft fell into the Department of Science and Culture. The Head of the Department A.M. Kutepov and his colleague G.T. Voronov came to Chernogolovka, got acquainted with our work and said: "Let's make another resolution. Your discovery is much wider than it was shown in the draft" (it concerned only the technology of hard alloys). G.T. Voronov prepared a new resolution and after my report at the meeting of the Council of Ministers it was accepted.

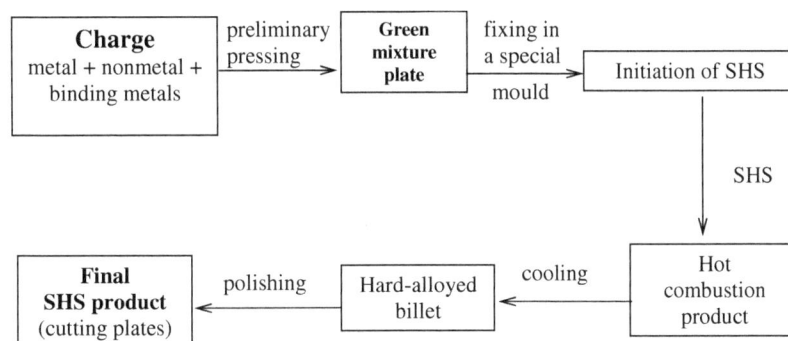

Fig. (5.1.1): Technological scheme of direct SHS production of tungsten-free hard alloys (I.P. Borovinskaya, V.I. Ratnikov, A.G. Merzhanov, A.N. Pityulin).

Due to the resolution the Scientific Council of SHS theory and practice was organized. We had good contacts with the Chairman of the USSR State Committee on Science and Technology academician G.I. Marchuk who was a purposeful, friendly person appreciating everything new. A special unit was also projected its construction was started in Chernogolovka.

The tasks to be solved:

- Simplification of combustion product machining to cutting plates

- Search for the most efficient application fields

- Analysis of economic efficiency

As to Ministries and Departments, they had different attitude to the new technology. For instance, we established good business relations with the Ministry of aviation industry at once and started our cooperation. But the specialists in hard alloys and tool materials (the Ministry of tool-and-die and machine-construction industries and the Ministry of non-ferrous metallurgy) did not want to apply our elaboration in spite of its advantages.

A.N. Pityulin (**head of the STIM project**)

Material	Density, g/cm^3	Average particle size, μm	Hardness, HRA	Bending strength, MPa	Application
STIM-1B/3	4,94	5–7	93,5	700–800	Unreground cutting plates
STIM-2A	6,40	1–2	87	1600–1800	Punches, drawing plates
STIM-3B/3	5,37	3–4	92,5	800–1000	Unreground cutting plates
STIM-3B	5,40	2–4	92,5	700–800	Nonscaling items
STIM-4	4,20	1–2	86	700–800	Heat-resistant items
STIM-5	5,70	1–2	90	1000–1200	Unreground cutting plates

Fig. (5.1.2): Synthetic hard tool materials (STIM) – a new type of hard alloys (A.N. Pityulin, I.P. Borovinskaya, A.G. Merzhanov).

We began working on the state level and were plunged into the world of political collisions where unfortunately, the common sense was not the main thing.

Our Council started to work actively. It included energetic, keen people who developed the working plan. We were discussing the results, accepting current decisions. I had to make annual reports about our scientific-and-technical results and the work of the Council.

Leaders of Ministries and Departments took part in our annual sessions. Every year we organized one or two thematic meetings in various towns. They were always a success.

By that time our team in Chernogolovka had extended the experimental investigation, developed a set of technological approaches (Figs. **5.1.3** and **5.1.4**) which are used now for pilot and semi-industrial production.

This period is remarkable for wonderful work of Dr. V.L. Kvanin in SHS compaction of hard-alloyed large-sized items. The work is very difficult for many reasons. The main one is: it is impossible to make many experiments with such considerable green mixture consumption! It is necessary to think more. Among the results achieved by V.L. Kvanin and his colleague N.T. Balikhina I'd like to point out the development of wear-resistant rolls, stop valves and other custom-made items (Fig. **5.1.5**) [25, 27].

V.L. Kvanin is a very talented researcher with an application tendency, an excellent inventor. He always has a lot of ideas. It's good. But sometimes his new idea surpasses the previous one and does not allow him to realize the started work.

Some experience gained during the initial years proved that our pilot SHS technologies met the requirements of semi-industrial production due to their high output. It became clear that technological lines of high automation level are necessary only for large-tonnage production. It made our work in the technology assimilation much easier.

Initialcomponents	**Powders:** metals, non-metals, oxides, hydrides, inert additions, regulating admixtures
	Gases: nitrogen, hydrogen, oxygen, inert gases, gaseous compounds

Preparation of initial components for synthesis :
·drying
·weighing
·mixing
·granulation
·shaping

solid + solid (powder mixtures)
solid + gas (hybrid systems)

Initialsystem	

SHS:
·in the open air,
·in reactors, in thermal vacuum chambers, in special constant pressure vessels,
·moulds,
·extruders,
·centrifuges,
·under external forced impacts and without them

Combustion product	

Low and hard sintered products: cakes, ingots, details, and items
Processing of products:
milling, chemical and mechanical disintegration, classification, surface finishing (grinding, polishing), dressing

Inorganic compounds: borides, nitrides, carbides, silicides, intermetallics, oxides, chalcogenides, phosphides, etc.
Reduced metals and compounds: (Ti, Mo, W, etc.)
Multicomponent materials: ceramics, cermets, mineral ceramics, composites
Articles

Final product	

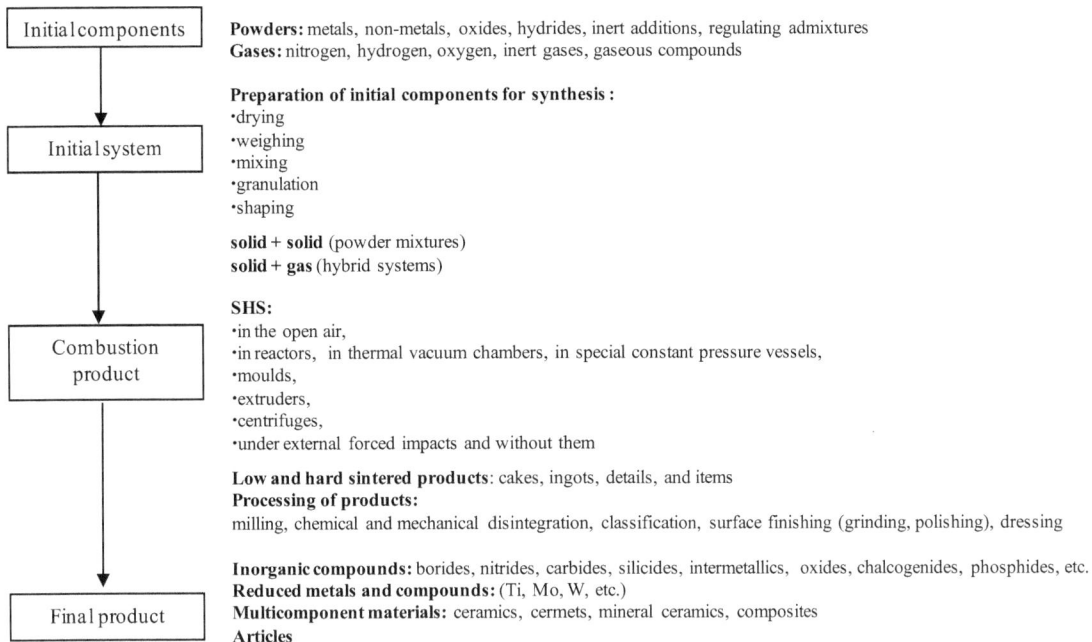

Fig. (5.1.3): Common scheme of three-stage SHS technology.

SIX MAIN TECHNOLOGICAL TYPES (TT) OF SHS

A.G. Merzhanov (beginning of the 80th)

TT-1		Technology of powder
MTI	–	SHS organization under the terms yielding the product of the preset
P	–	powder mixtures "solid+solid" , hybrid systems "solid+gas",
MT	–	obtaining of billets of preset composition for powder production
A	–	raw materials for powder production
TT-2		SHS sintering
MTI	–	reagent shaping and organization of SHS under the terms keeping the sample shape and size,
MT	–	synthesis of porous items,
P	–	SHS in P – SHS in gasostats and thermal-vacuum,
A	–	filters, catalyst carriers, billets for impregnation with metals
TT-3		Forced SHS compaction
MTI	–	densification to non-porous state of a hot product by external forces,
MT	–	synthesis of non-porous materials,
P	–	densification methods: pressing, all-round compression, shock and shock-wave loading, extrusion, rolling,
A	–	billets and items of tungsten-free hard alloys
TT-4		Technology of high-temperature SHS melts or SHS metallurgy
MTI	–	selection of highly caloric mixtures yielding molten products at combustion,
MT	–	synthesis of cast materials,
P	–	organization of the process in hermetically sealed reactors or centrifuges,
A	–	metal pipes with an internal ceramic layer for abrasive media transportation, items with protective wear-resistant coatings
TT-5		SHS welding
MTI	–	SHS organization between two refractory conducting items with additional heat release
MT	–	permanent connection (welding) of hardly welded items
A	–	welding of metal items and graphite

TT-6		SHS technology of gas-transport coatings
MTI	–	organization of SHS in the same way as TT-1 but with additional gas-transport reagents and thin coating of the item with synthesized product,
MT	–	application of thin wear-resistant coatings,
A	–	application of thin protective films on cutting tools

Fig. (5.1.4): Classification of SHS-technologies: MTI – main technological idea, MT – main target, P – procedure, A – application [4].

Fig. (5.1.5): Large-scaled items and the leading specialist in this field Dr. V.L. Kvanin.

5.2. "TERMOSYNTEZ"

In 1986 a new State Resolution was issued on organization of interbranch scientific-and-technical units (MNTK in Russian). They should have been informal organizations solving one and the same problem. The aim of their creation was quite clear – to accelerate the development of a scientific elaboration up to its practical application. The problem was to be solved by concentration of efforts of various institutions working on the same subject. The authors realized that it was too long to create a new complex organization. But acceleration of scientific-and-technical progress was a part of our state policy.

Sixteen considerable problems were chosen. They were expected to be of benefit for our state economy.

We were very glad to learn that our unit had been included in one of those sixteen. It was called MNTK "Termosyntez". I think it was an echo of the initiative activity of our Scientific Council. We started working on organization of our MNTK. It was subordinate to both the Academy of Sciences and the Ministry of non-ferrous metallurgy. At first I got upset but then I realized the advantage of such a solution: our enemies became our friends. We began working friendly indeed. We really made a lot during the years of MNTK existence (1987–1997).

During those years our team in Chernogolovka worked within the department of macroscopic kinetics of the sector of macrokinetics and gasodynamics.

The Institute of Chemical Physics and its Branch in Chernogolovka were divided into seven large sectors. They were to develop the most important directions and problems of chemical physics. Our sector was the youngest. It was created by the Director of the Institute N.N. Semenov under the influence of our progress in SHS investigations. It consisted of two subsectors: macrokinetic and gasodynamic ones. The sector of gasodynamics was headed by my colleague Prof. A.N. Dremin, an outstanding specialist in the field of explosive detonation. Our SHS group had "grown up" by the time: some new laboratories had been organized. Chemical and chemical-engineering investigations were concentrated in the Laboratory of SHS problems, which was headed by Prof. I.P. Borovinskaya. Macroscopic studies of SHS processes were carried out in several laboratories. In the resolution on "Termosyntez" organization we could read that the head organization of the unit was the Department of Macroscopic Kinetics of the Institute of Chemical Physics of USSR AS. It was a department, not an institute! The main inspirers of the organization of those units thought that SHS could be lost in a huge institution. But the fact that a powerless department (not a powerful institute) headed a large organization was nonsense. The situation was unstable and soon it was broken down. Our department was reorganized to the Institute which became the Head organization of the complex.

By that time the new facilities had been built in Chernogolovka and it became the residence of our new Institute. All the colleagues of our Department went to the new Institute except a small group. We decided that organization of some new laboratories would result in fast development of new investigation and possibilities for young specialists to unveil their potential. I.P. Borovinskaya had the largest laboratory. She suggested that some of her colleagues who had promising directions should start working independently. Thus, new laboratories appeared. They were headed by S.S. Mamyan (magnesium-reduced SHS processes), A.N. Pityulin (forced SHS compaction), L.V. Kustova (chemical analysis) and later by A.M. Stolin (rheology, rheodynamics, and SHS extrusion), V.I. Ponomarev (X-ray analysis and crystallography), V.I. Yukhvid (aluminum-reduced SHS-processes), V.I. Ratnikov (SHS-equipment), G.A. Vishnyakova (physical materials science).

It was a great pleasure for me to design the structure of the Institute. At first it was necessary to name our Institute. I did not want the Institute to be called the Institute of self-propagating high-temperature synthesis. Certainly, it was SHS that brought us to the Institute foundation but it could not provide us with an extended scope of investigations. I did not want to use ordinary words "macroscopic kinetics" to name the Institute. So after long discussions and disputes O.E. Kashireninov (the Deputy Director of MNTK) and I decided to name it the Institute of Structural Macrokinetics. There were no Institutes with such a name.

But it was a formal side of the problem. I really wanted to extend the research themes but preserving their ideology. I think we were a success. We found some adjacent research directions, organized new laboratories, invited some scientists to head them (Fig. **5.2.1**).

At the same time the Council of Ministers issued the next resolution on creating several institutions in different regions which were destined to assist ISMAN in running the work and to extend the scope of SHS investigations. Among them was Kazakh Interbranch Research SHS Center in Alma-Ata and Georgian Research Center at the Institute of Metallurgy of the Georgian Academy of Sciences.

Soon a Branch of our Institute in Tomsk and the SHS Center ISMAN-MISiS were organized. Thus, a large powerful group appeared to solve research and technological problems of SHS. So we acquired a taste of real work due to the State support.

I.P. Borovinskaya

V.M. Shkiro

B.I. Khaikin

A.P. Aldushin

Yu. M. Grigoriev

V.I. Ratnikov   **V.K. Prokudina**   **G.A. Vishnyakova**

A.K. Filonenko   **V.M. Maslov**

Fig. (5.2.1): Pioneers of SHS.

5.3. ON ISMAN AND TERMOSINTEZ ACTIVITY IN 1987 – 1992

It was clear that MNTK was created to advance the scientific elaborations and introduce them in industry. We had had some unrealized elaborations so we should have start with them and thereby to bear out our potentialities.

We were responsible for the work. We selected our more advanced elaborations suitable for industrial assimilation, got in touch with the members of Ministries and State Planning Committee. At the same time we organized some departments in the Institute which were necessary for presiding the MNTK (groups of marketing, technical and economical efficiency, standardization, technical information). Also we made a unified plan of the MNTK "Termosyntez" where we reflected all our elaborations and plans.

The results of our work are shown in Fig. **5.3.1**. You can see semi-industrial facilities in various plants of the country. The main direction was production of inorganic powders by powder technology (it is still the most developed SHS technology). But also we can see the production of items (cutting plates, ceramic insulators for furnaces of oriented crystallization) **[28]**.

The total output of SHS products in 1992 was 2000 tons.

We made sure that SHS could be used for solving industrial tasks. But we realized that the success was predetermined by our previous work when we acted unhurriedly and scrupulously and tried to prove that SHS could serve the purpose of the scientific-and technical progress. That is why in order to make a reserve for our future innovations we organized some new research directions of the applied character.

During 5 years within the MNTK "Termosyntez" we realized more than 100 elaborations. Some of them are listed in Fig. **5.3.2**.

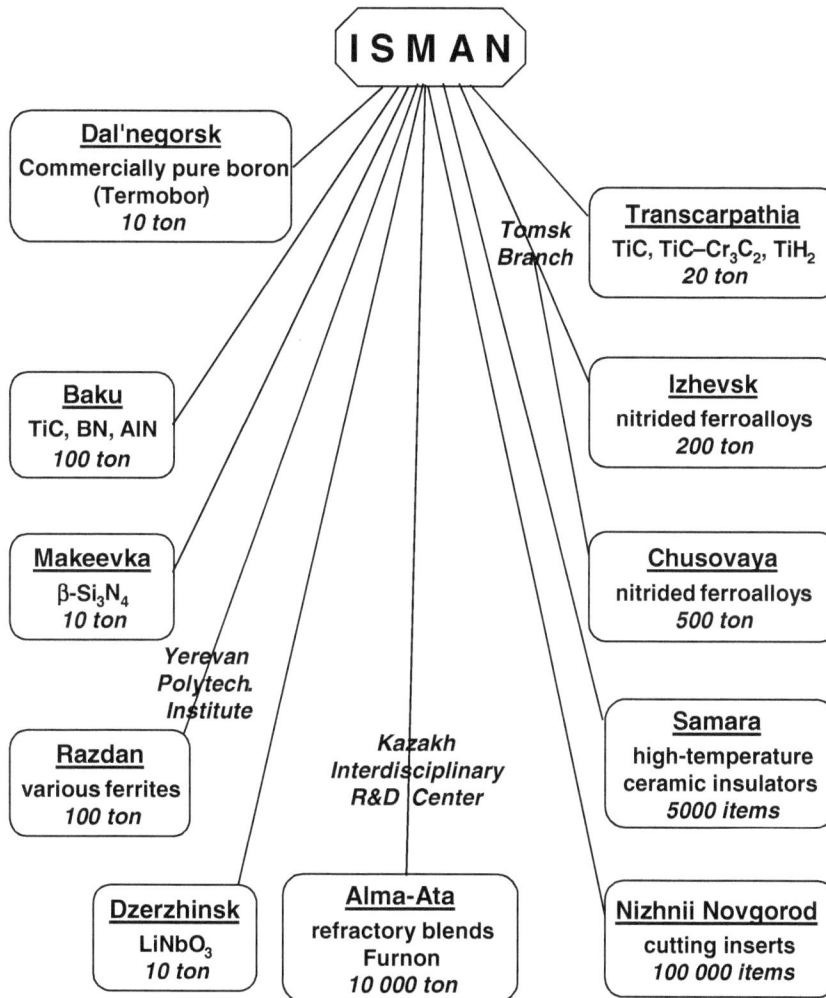

Fig. (5.3.1): Industrial application of SHS in the former USSR.

Powders, Materials, Items, Coatings	Technology and Equipment
1. SHS powders of refractory compounds of high purity: • TiC and TiCN for abrasives, coatings, steels and hard alloys; • Si_3N_4, AlN for ceramics and fillers of composite materials; • MoS_2 ($MoSe_2$), WS_2 (WSe_2) for solid lubricants.	1. Continuous SHS technology for production of Ni-Zn and Mn-Zn ferrites of up to 1000 tons/year output
2. SHS powders of silicon and aluminum nitrides with preset properties	2. SHS technology with a reduction stage for obtaining: • SHS fine (0.1 μm) powders of B4C, BN, TiB2, TiC, amorphous boron; • SHS cast powders for surfacing including Cr3C2+Al2O3·Cr2O3.
3. Materials and items obtained by SHS densification and used under the terms of intensive wear, friction and aggressive chemical media: • STIM-C plates; • STIM-5 unreground cutting plates; • Large-scaled rings for rolls; • Plates of functionally graded materials.	3. SHS-azide technology for obtaining refractory powders including composite ones. It is an alternative technology due to utilization of halogen salts (usually wastes of chemical production), oxides of the elements to be nitrided, and sodium azide (previously – an initial raw material for obtaining explosives).

4. Porous materials and items • Based on SHS titanium nickelide for medicine, • Ceramic and cermet filters for water, oil, and gas suspension	4. Obtaining of light SHS blocks used for lining of furnaces operating at T up to 2300 K in aggressive media.
5. Ceramics obtained in constant pressure vessels (SHS gasostats) at high pressures (up to 300 MPa) and used at high temperatures and in aggressive media: • BN bushings, • Stop valves, • Equipment for single crystal growing, • Triggers for stomatology.	5. Small plants of water and oil products using SHS ceramic filters. The efficiency is 20 m³/h, operation life – 10 years.
6. Electrodes obtained by SHS extrusion and SHS casting for applying protective coatings on surfaces of metal items and tools by electric-spark alloying, electric-arc, plasma and induction surfacing and plasma spraying	6. Reaction moulds for SHS pressing, i.e. combustion with simultaneous densification of hot combustion products.
7. Refractory materials	7. SHS gasostat (constant pressure vessel) for obtaining materials and items by the SHS method at P = up to 500 MPa.
8. SHS pigments (more than 150 tints)	8. Universal SHS equipment for synthesizing SHS materials as cakes, ingots and powders.
9. SHS mixture used for growing single crystals of lithium niobate for electronics	9. Fast calorimeter used for studying physical and chemical processes occurring during SHS and following with heat release.
10. Disks, electrodes, targets for surface hardening of tools.	10. Equipment for time-resolved XRD analysis

Fig. (5.3.2): Some achievements of MNTK "Termosyntez" in 1987 – 1992.

Let's dwell on one example – SHS of complex oxides. They include niobates, titanates, silicates, ferrites, cuprates, etc. They are widely used. There is no industry or technical construction which can do without these materials.

There are a lot of ways for synthesizing complex oxides. The most known of them is thermal homogenizing of powder mixtures of simple oxides **[29]**.

$$\sum_{i=1}^{k}\left(Me_{m_i}O_{n_i}\right)_i + Q = \prod_{i=1}^{k}\left(Me_{m_i}O_{n_i}\right)_i$$

Also there are some other SHS variations for obtaining complex oxides.

But in 1977 an "elegant" method was proposed by V.V. Boldyrev. It is a direct synthesis from compounds (simple oxides)

$$PbO_2 + WO_2 \rightarrow PbWO_4 + Q$$

However, because of the low values of the reaction heat evolution for most of the mixtures it could not find any practical application.

In the MNTK "Termosyntez" M.D. Nersesyan and I.P. Borovinskaya proposed another method. In the mixture of simple oxides which was used in the above mentioned method one of the oxides with rather high heat formation is substituted for the same metal. The mixture becomes combustible and is ignited in the air or oxygen.

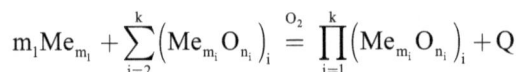

$$m_1 Me_{m_1} + \sum_{i=2}^{k}\left(Me_{m_i}O_{n_i}\right)_i \overset{O_2}{=} \prod_{i=1}^{k}\left(Me_{m_i}O_{n_i}\right)_i + Q$$

The metal burns out yielding oxide and warms up the mixture up to high temperatures. And if the product cooling is not very fast, its autoannealing occurs and there is enough time for homogenizing to be over and yield to complex oxide formation (the cooling time can be chosen experimentally). The efficiency of the method is shown in Fig. **5.3.3** ($YBa_2Cu_3O_{7-x}$ synthesis). This compound was synthesized in 1987 after Muller and Bednors's discovery of high-temperature superconductivity. That time the compound was characterized by the highest transition temperature ~93 K. However, the compound is known to be very uncertain. It easily releases oxygen (always x>0) and can lose its superconducting properties.

Fig. **5.3.3** shows the comparison of some characteristics of SHS powders obtained in ISMAN (SHS J1 and J2) with those of the powders syntsized in furnaces in American Companies (SC5-S – SSCO3-0065). According to the opinion of the main experts who tested our products in the USA our powder SHS J2 was one of the best all over the world. In the 80-s ISMAN delivered these powders to the American Company "HiTc Conco" which used them for sintering high-dense items. Development of these powders is a science-intensive elaboration. We had to study thoroughly the chemical reaction mechanism in the combustion wave. Then in the 90-s all the known superconductors were synthesized in ISMAN **[30]**.

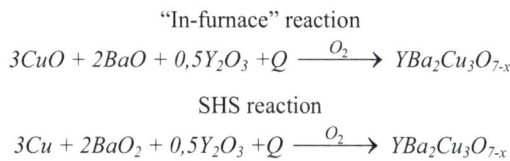

"In-furnace" reaction

$$3CuO + 2BaO + 0,5Y_2O_3 + Q \xrightarrow{O_2} YBa_2Cu_3O_{7-x}$$

SHS reaction

$$3Cu + 2BaO_2 + 0,5Y_2O_3 + Q \xrightarrow{O_2} YBa_2Cu_3O_{7-x}$$

Combustion wave   **Microstructure of the product**

Sample	Oxygen content, at. Units (7-x)	Critical temperature, Tc, K	Ortho-phase content, Y123, %	Main impurities	Average particles size, μm
SC5-S	6.92	93.5	98	CuO	40.0
SC5-P	6.87	93.5	~99	CuO	8.2
SC5-6.5	6.88	90.5	97	BaCuO₂	6.8
SC5-7	6.85	92.5	~99	CuO	5.7
CPS A-1203	6.85	92.0	~99	BaCuO₂	3.0
SSC 03-0065	6.89	92.0	~99	-	6.5
SHS-J1	6.90	92.0	97	CuO	9.0
SHS-J2	6.92	93.5	~99	-	8.0

Fig. (5.3.3): Synthesis of high-temperature superconductors $Ba_2Cu_3O_{7-x}$.

All our specialists started their work at the Institute of Chemical Physics. It was a powerful scientific organization headed by a Noble Prize winner academician N.N. Semenov. He managed to create in the Institute the atmosphere of keenness, creative impulse, aspiration for discoveries and understanding of the results. N.N. Semenov and then F.I. Dubovitsky inculcated in us the spirit of harmonious creative work combining the interest both to fundamental and applied investigations and a great wish to use the results in practice. That is why when we agreed to participate in the MNTK with the main aim to apply the scientific results in practical work we could not be engaged only in fundamental research.

But first of all I wanted to know the essence of "structural macrokinetics". I wanted to know what the ideology was and what method could be used in structural macrokinetics. Now I'd like to cite some conclusions from one of the papers:

- "Phase and structure transformations and their connection with chemical reaction kinetics and heat- and mass-transfer are of great importance for structural macrokinetics. SHS theory must be based on the concepts of structural macrokinetics.

- There are two investigation approaches: structural statics and structural dynamics. Structural statics deals with final product characteristics (composition, crystallite dimensions, crystal lattice parameters, porosity, etc.) and their relationship with the process and green mixture parameters (composition, particle size, green mixture density, its temperature, gas pressure, etc.). This approach can be easily realized in the experiment but it is rather difficult for understanding. Structural dynamics is connected with mechanism and dynamics (kinetics) of the product structure formation, i.e. the substance transformations within the process. In this case it is difficult to realize the approach in the experiment but it is much easier to understand and explain the results.

- Structural statics allows studying the characteristic-parameter dependences as well as discovering and describing new effects.

- The efficient methods in structural macrokinetics are the method of dynamic XRD analysis and sample annealing in a slow combustion wedge.

- The approaches developed in structural macrokinetics can be used not only in SHS but in all material-forming chemical processes – such as plasma-chemical synthesis, chemical-condensation processes (gas reactions with the product condensation, etc.).

Some results are presented in Figs. **5.3.4– 5.3.9. [31].**

Example 1:

The temperature profile of a "classical" combustion wave (Zel'dovich, Frank-Kamenetsii, 1934)–classical macrokinetics

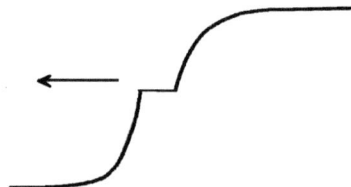

The temperature profile of a combustion wave where phase transformations take place (Merzhanov, 1973) – structural macrokinetics

Example 2:

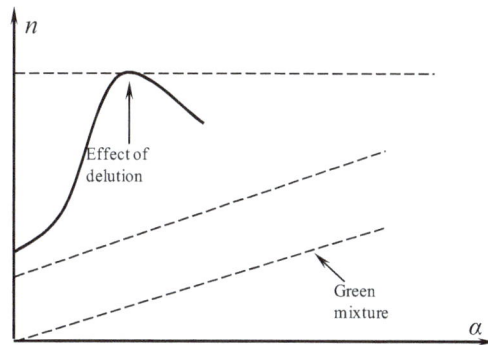

Fig. (5.3.4): Ideology of structural macrokinetics (I) SHS = combustion + structure formation.

SHS wave structure in the case of equilibrium mechanism
(Khaikin – Aldushin – Merzhanov)

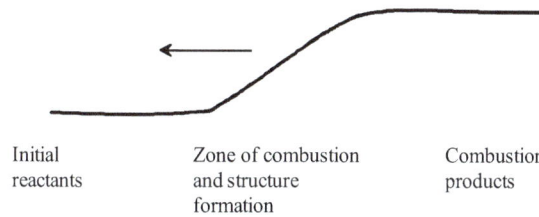

Initial reactants	Zone of combustion and structure formation	Combustion products

Combustion and structure formation occur in one zone not far from the front and have an influence on combustion velocity

SHS wave structure in the case of non-equilibrium mechanism (Borovinskaya)

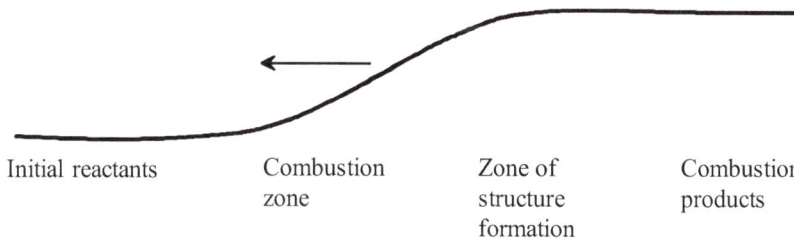

Initial reactants	Combustion zone	Zone of structure formation	Combustion products

Structure formation takes place far beyond the front, and the combustion velocity depends on the reaction rate (not on structure formation)

Fig. (5.3.5): Ideology of structural macrokinetics (II). (See also Fig. **3.10.**)

Combustion of Ti + Si system in the cooling copper block with a wedge gap

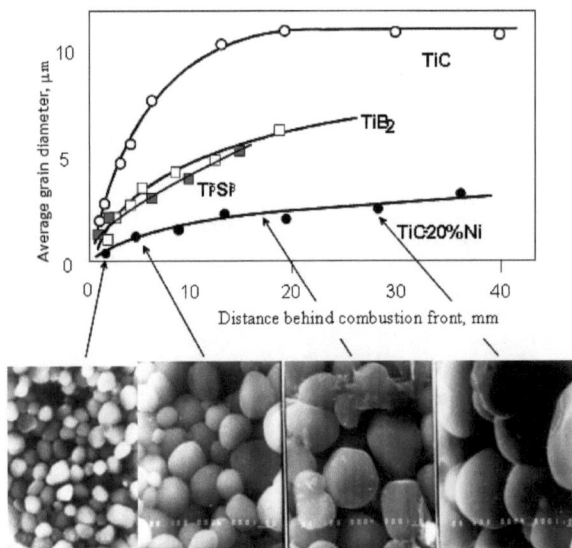

TiC grain growth after front propagation for 80% (TiC+C) + 20% Ni system

Fig. (5.3.6): Methods of Structural Macrokinetics I. Hardening of SHS wave in wedge-shaped samples placed in a huge copper block for studying microstructure evolution in the combustion wave (by A.S. Rogachev) [32, 33].

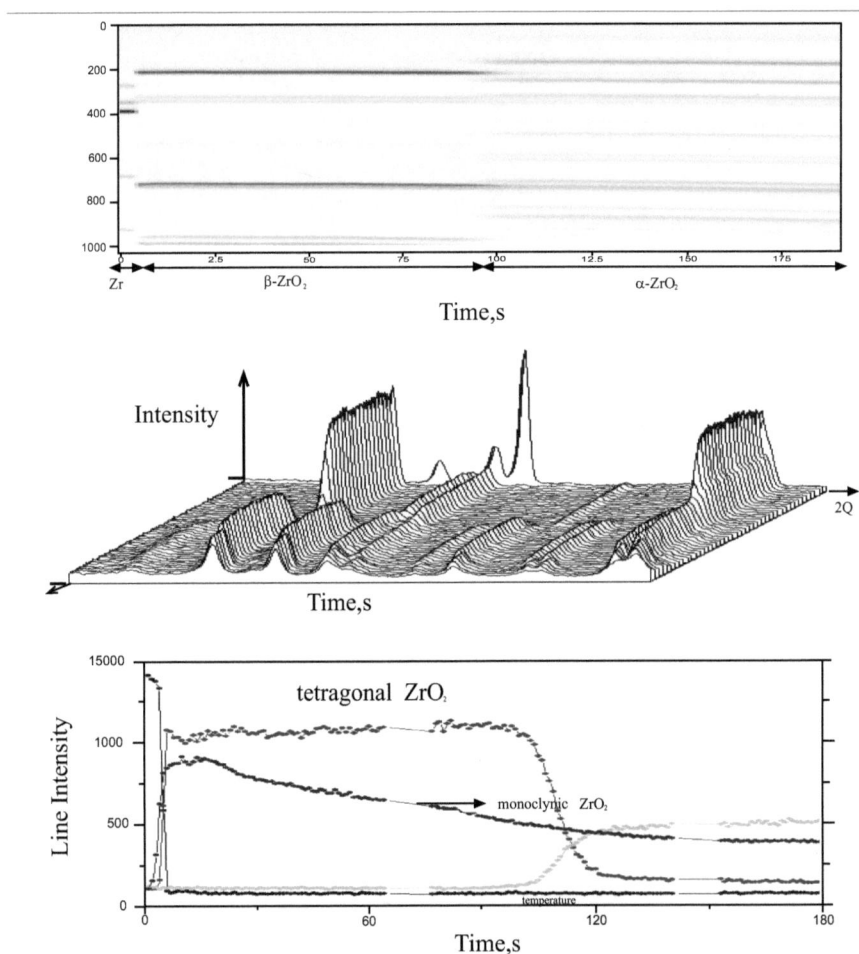

Fig. (5.3.7): Methods of Structural Macrokinetics II. Time-resolved diffraction patterns of the product formed upon Zr combustion in air, 3-D picture of the process (by D.Yu. Kovalev, V.I. Ponomarev, V.M. Shkiro) [34, 35].

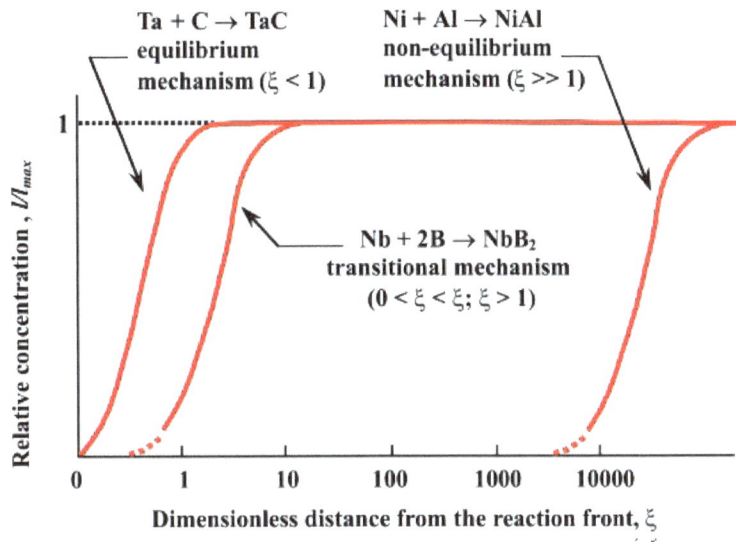

Initial mixture	Intermediate phases	Final product
Ti + air	TiN, Ti_2O_{5-x}, $TiO_{2-x}N_x$	TiO_2
Ni + Al	Were not found out	NiAl
Nb + 2B	-	NbB_2
Nb + B	NbB_2	NbB
2Ta + C	-	Ta_2C
Ta + C	Ta_2C	TaC
Ti + C	-	TiC

Fig. (5.3.8): Methods of Structural Macrokinetics III. Mechanism of phase formation as derived from TRXRD data. The diagram – A.G. Merzhanov, 1993; the Table – I.P. Borovinskaya, A.G. Merzhanov, I.O. Khomenko, A.S. Mukasyan, V.M. Shkiro, V.I. Ponomarev, 1993.

SHS ferrite obtained by the reaction
$$12Fe + BaO_2 \xrightarrow{O_2} BaO \cdot 6Fe_2O_3,$$
and characterized by spontaneous magnetization (P.B. Avakyan, 1989).

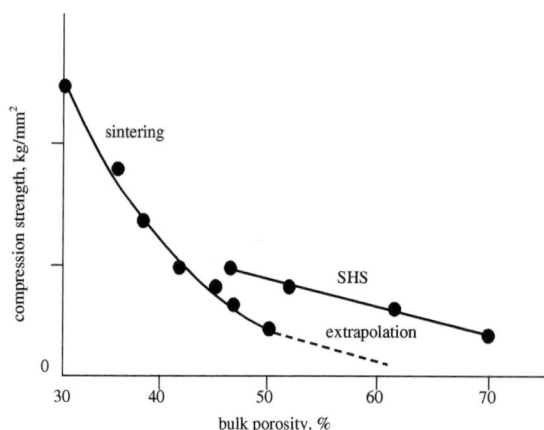

Porous titanium carbide obtained by SHS is stronger than a sintered analog due to strong bonds between grains (V.N. Bloshenko, 1990).

Fig. (5.3.9): New results in structural statics [36].

5.4. "TERMOSINTEZ" RESEARCH CENTERS

According to the Resolution of the Council of Ministers some research and technological centers were organized for enlarging the scope of SHS work.

Kazakh SHS Research Center in Alma-Ata started its activity under the leadership of G.I. Ksandopulo who was the Head of the Department of Kinetics and Combustion in Kazakh University (Fig. **5.4.1**). We have been friends for a long time and organized a lot of joint events in Alma-Ata.

When the Center was organized, we gave them our elaboration of refractory synthesis. When using the reagents of the reaction purity (we did not have any others), we got very good results. But it was clear that a refractory couldn't be made of chemical reagents – it was too expensive. There were a lot of mineral raw materials in Kazakhstan and the aim of producing such materials there was rather promising.

G.I. Ksandopulo

Director of Kazakh Research Center of "Termosyntez"

Main investigation subjects of the Center:
Oxide materials and substances
- Refractories
- Pigments
- Catalysts

The first International Symposium on SHS was held in Alma-Ata in 1991.

Fig. (5.4.1): Kazakh SHS Research Center of "Termosyntez".

G.I. Ksandopulo was very active in his work and soon he achieved excellent results. His refractory materials had high operation characteristics and were very cheap. His colleagues created various types of the materials (their total name was "Furnon"), and developed the scheme according to which the main product was not the material but the green mixture (and a customer would make the refractory material himself in the place where it was necessary). Production of refractory materials was the main subject under study in the Center (Fig. **5.4.2**) [37]. But G.I. Ksandopulo branched out their investigation theme. One of the active researchers was Galina Ksandopulo, a daughter of G.I. Ksandopulo. She is a young, beautiful, clever and purposeful woman. Now she lives in Greece (Fig. **5.4.3**).

But the SHS Center did not exist long. It was reorganized to the Institute of Combustion Problems and taken out of our MNTK. But we have preserved our creative links.

Technical characteristics of "Furnon" refractories
High strength of brick work (up to 15 MPa) at $T_{oper.}$= 1400°C

Destination
In shaft furnaces, rotary kilns for limestone, cermet, coke, and anode past kilns,
for crowns of melting furnaces.

Existing analog
Refractory mortars – alumosilicate, periclase, chromite-periclase

Effectiveness in comparison with the analog
Increase of overhaul life of furnaces, decrease of capital and labor expenditures by 40–70 %

Fig. (5.4.2): SHS – refractories (Brand name «Furnon»). G.I. Ksandopulo, M.B. Ismailov, *et al.*, 1990.

G. Xanthopoulou

Fig. (5.4.3). SHS pigments for plastic and glass.

Georgian SHS Research Center in Tbilisi was organized in the Institute of Metallurgy of the Georgian Academy of Sciences. The Institute was headed by academician F.N. Tavadze – a creative friendly and brave person. Those days our Soviet metallurgists did not like SHS, they said that they had known everything before (Why hadn't they developed it?). But F.N. Tavadze did not listen to their critics and joined us. We had begun our cooperation with two researchers of the Institute G.F. Tavadze and G.Sh. Oniashvili before the Center organization. We liked that they were fascinated by the SHS and were eager to develop SHS processes and materials in Georgia. When we discussed the work with F.N. Tavadze, we concluded that it was interesting to carry out some metallurgical processes by SHS. For example, they produced amorphous alloys as a band (by hardening the melt injected on the revolving drum). It was clear that the melts could be obtained by SHS. Another direction was connected with the use of local manganese ores for synthesizing manganese-containing ferroalloys. And again we accepted some talented young people for employment, and they came to Chernogolovka to study.

The young people were very serious and enthusiastic. When working in our Institute, they defended their theses. G.F. Tavadze and G.Sh. Oniashvili (they have the same name Georgy and we called them two Georgies) often came to Chernogolovka, they took care of their young colleagues. But at the same time the work was being developed in the Institute in Tbilisi. Two Georgies found researchers who were interested in SHS and a new enthusiastic group was born in Georgia.

When the young SHS specialists came back from Chernogolovka to Tbilisi, the themes of investigation of the Georgian SHS Center were expanded. G. Oniashvili became interested in intermetallics synthesis, and

G. Tavadze was engaged in obtaining hard alloys with phosphate binders by the method of forced SHS compaction.

One of the most developed directions in the Institute of Metallurgy was connected with boron and its compounds. They managed to synthesize borides based on radioactive boron isotope (B^{10}) by the SHS method. Our Georgian colleagues are open for cooperation. They like their work and are eager for studying. Not long ago G.F. Tavadze became Director of the Institute of Metallurgy named after F.N. Tavadze. I hope that his successful work in SHS helped him.

You can see the information about the Georgian SHS Center in Figs. **5.4.4–5.4.5 [38 41].**

G.F. Tavadze   G.Sh. Oniashvili

Georgian SHS Research Center of "Termosyntez"
at the Institute of Metallurgy of the Georgian Academy of Sciences

Main investigation subjects:

1. SHS analogs of some processes in ferrous and nonferrous metallurgy.
2. SHS application in some stages of conventional metallurgical processes (obtaining of amorphous bands from SHS melts).
3. SHS processing of local ore and mineral materials.

Fig. (5.4.4): Georgian SHS Research Center of "Termosyntez".

Production of carbon-free ferromanganese with high manganese content in the ingot (>85 %) for low carbon steels and special carbon-free alloys.

Efficiency – development of the technology of a cheaper product.
Authors: G.Sh. Oniashvili, Z.G. Aslazashvili, et.al., 1989.

Production of single-phase alloys based on TiAl (Ti$_3$Al, TiAl, TiAl$_3$)

Main technical characteristics:
yield point – 1210 MPa
breaking point – 1890 MPa
relative elongation – 20 %

Fig. (5.4.5): Some achievements of the Georgian SHS Research Center.

SHS ENGINEERING CENTER IN SAMARA POLYTECHNICAL UNIVERSIT

We had established friendly relations with Kuibyshev Polytechnical Institute (now it is Samara Polytechnical University) long before we started our investigation of SHS. We organized the department of engineering chemical physics in our section and their students came to us for studying. Our scientists delivered lectures in chemical physics, macrokinetics, and combustion theory. When we started our work in SHS, we tried to attract the students' interest to this problem. When our former student A.P. Amosov became interested in SHS, our work was being developed actively. A.P. Amosov is a clever, easy-tempered, highly organized man. He graduated from our department, continued his study and became a specialist in explosive's sensitivity to mechanical actions. His first work in SHS was connected with SHS mixture ignition under the terms of impact and friction actions. Then A.P. Amosov organized the SHS Engineering Center which was included in MNTK.

The colleagues of the Center are interested in the development of nitrogen-containing material technologies based on the use of azides as starting components (Figs. **5.4.6–5.4.8**).

Main directions of R&D activity of the Engineering SHS Center in Samara University

1. Azide-based technology of SHS ceramic nitrogen-containing powders
2. Filtration technology of ceramic and composite powders
3. Forced SHS compaction of hard alloys
4. SHS alloying elements
5. Sensitivity of SHS composites to mechanical effects.

A.P. Amosov, Director of the Center

Fig. (5.4.6): Engineering SHS Center in Samara University.

Azide-based technology (SHS-Az)
Authors – V.T. Kosolapov, A.G. Merzhanov, 1978.
The work was continued by A.P. Amosov, G.V. Bichurov.

Advantages:

1. Utilization of "internal" nitrogen
2. Separation of reactive mass and combustion products due to abundant gas release
3. High-dispersion of the product due to absence of recrystallization processes because of low combustion temperature
4. Absence of sintering activators

Azide-based technology for obtaining micro- and nano-sized powders.

Reactions:

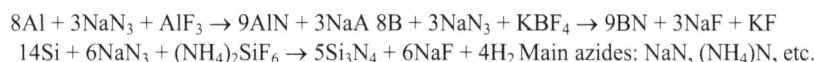

$$8Al + 3NaN_3 + AlF_3 \rightarrow 9AlN + 3NaA \quad 8B + 3NaN_3 + KBF_4 \rightarrow 9BN + 3NaF + KF$$
$$14Si + 6NaN_3 + (NH_4)_2SiF_6 \rightarrow 5Si_3N_4 + 6NaF + 4H_2 \quad \text{Main azides: NaN, } (NH_4)N, \text{ etc.}$$

Fig. (5.4.7): Azide-based technology developed in the Engineering SHS Center in Samara.

Filtration SHS technology

A.P. Amosov, A.G. Makarenko (1998); V.I. Nikitin, A.G. Makarenko (beginning of the 80-s)

Advantages:

1. Low pressure
2. Possible organization of low-exothermic reactions

Production of alloying elements for aluminum alloys
V.I. Nikitin, A.G. Makarenko
Green mixture Al + 5%Ti, Al + 10%Zr
(SHS occurs in the melt)

Advantages:

1. 1.5–3 times decrease in the components' size
2. Improvement of alloy properties
3. 10–20 % increase in bending strength
4. 2–4 times decrease in consumption of alloying additives
5. 5–10 % increase in output

Fig. (5.4.8): Some achievements of the Engineering SHS Center in Samara [42, 43].

SCIENTIFIC-AND-EDUCATIONAL CENTER SHS MISIS-ISMAN

The work in SHS was being developed actively but we did not have enough specialists in our field. MISiS (Moscow Institute of Steel and Alloys) was a leading educational institution with a metallurgical bias. Our MNTK was subordinate not only to the USSR Academy of Sciences but also to the Ministry of non-ferrous metallurgy. That is why we chose this institution. The creative role in organization of the Center belongs to N.N. Khavsky (MISiS) and I.P. Borovinskaya (ISMAN).

At the initial period they were scientific supervisors of the Center. The Director of the Center was the young specialist E.A. Levashov, a talented, energetic, and sociable man. He appeared to be not only a good scientist but also an excellent organizer. We acquainted E.A. Levashov with all our achievements and gave him the opportunity to choose the direction interesting for him. He started working in forced SHS compaction. The first work which was suggested by E.A. Levashov and N.N. Khavsky was investigation of ultra-sound effect on SHS compaction. It became an impulse for the further development of the Center.

However, E.A. Levashov realized that the main task of the Center was training of SHS specialists. This part of the Center's activity was carried out excellently too.

Now the Center is a leading team of scientists and teachers which gained the authority among the SHS specialists.

Some R&D achievements of the Center are shown in Figs. **5.4.9–5.4.11 [44, 45]**.

E.A. Levashov

Director of Scientific-and-educational Center SHS MISiS-ISMAN

Main Tasks:

1. Training of SHS specialists
2. Investigation of ultrasound effect on forced SHS compaction of hard alloys
3. Development of STIM 5 cutting alloy (with ISMAN)

Directions Under Study:

1. Theoretical combustion and structure formation models of various heterogeneous systems
2. A new class of composite electrode materials
3. Vacuum SHS rolling
4. Scientific backgrounds of thermoreaction electric-spark deposition of coatings
5. New classes of target cathodes
6. Types of magnetron spraying
7. Biologically active films
8. Diamond-based materials and tools
9. Mechanical activation in SHS

10. Porous TiAl-based filters for water purification

11. Ultrasound effect

12. Structure formation mechanism

13. Installation for electric-spark alloying

14. Investigation of ceramic materials made at ISMAN

Fig. (5.4.9): Scientific-and-educational Center SHS MISiS-ISMAN.

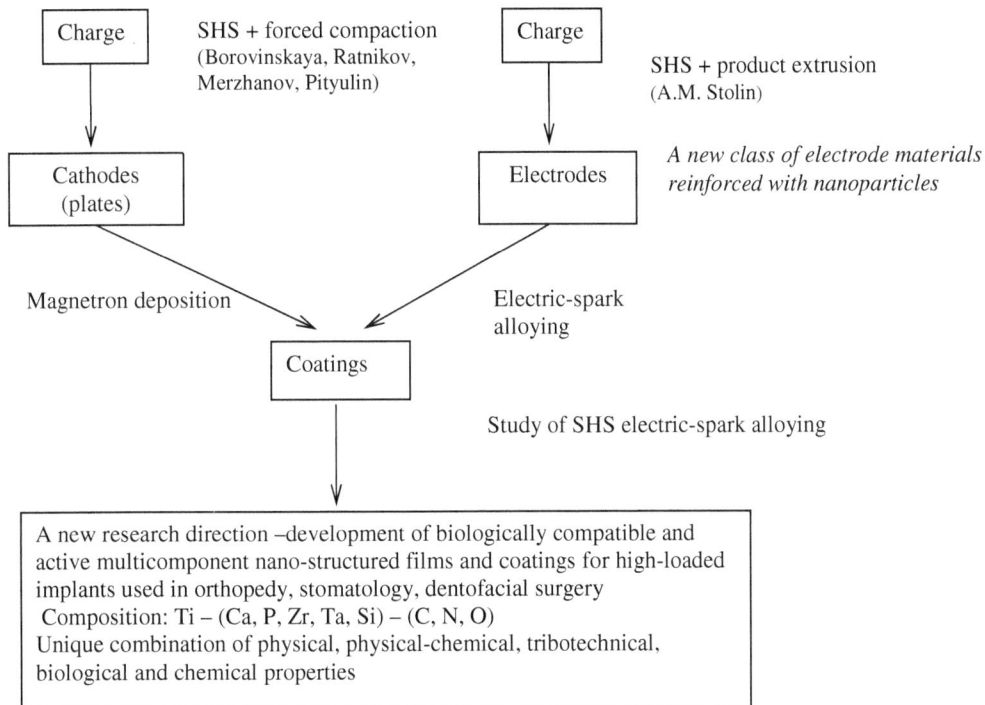

Fig. (5.4.10): Main directions of R&D activity of Scientific-and-educational Center SHS MISiS-ISMAN.

ISMAN BRANCH IN TOMSK

I have already written about the team headed by Yu.M. Maksimov in Tomsk. It was reorganized to the ISMAN Branch. It strengthened our friendship. The Branch was included in MNTK and the specialists made an important contribution to the development of SHS studing:

1. Generation of SHS fields in SHS processes

2. Mechanochemical synthesis of nano-sized powders

3. Mechanism of SHS processes forming intermediate melt in the combustion wave

4. Computer-assisted modeling of Macrokinetics of solid-phase chemical transformations under the terms of dynamic compression

5. Synthesis of large-scaled ceramic items

Now a few words about the collaboration with our Byelorussian partners. We had good links with the Institute of heat- and mass-transfer and especially with V.A. Borodulya who headed the Center of cooperation with socialist countries and held seminars on various problems of heat- and mass-transfer. These problems were of great interest for us as we developed thermal theory of self-inflammation, ignition and flame propagation at that time. As for me I supervised the part of heat- and mass-transfer in the systems reacting by a chemical route at these seminars.

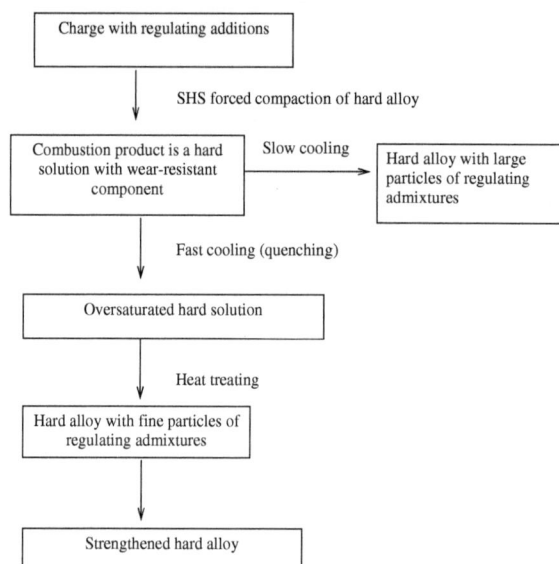

Fig. (5.4.11): Scientific-and-educational Center SHS MISiS-ISMAN: concept of formation of physical-mechanical properties of STIM hard alloys. E.A. Levashov, A.G. Merzhanov (2007).

Soon our Byelorussian colleagues became interested in SHS and joined us.

Now this scientific direction is headed by the Vise-President of the Byelorussian Academy of Sciences academician P.A. Vityaz. Some information about the results of our Byelorussian colleagues is presented in Fig. **5.4.12**.

The State support of our work was very important for us. It was the most prominent period in our activity. The number of scientists and organizations involved in our theme increased greatly.

The main role in the organization of this support belonged to our statesmen N.K. Baybakov, G.I. Marchuk, N.I. Ryzhkov, E.K. Ligachev, L. L. Voronin (responsible for MNTK activity). **[17-29]**.

Institute of heat and mass exchange of the Byelorussian Academy of Sciences	• Filtration combustion • Percolation phenomena in heterogeneous systems • SHS in multilayer films
Physico-technical Institute of the Byelorussian Academy of Sciences	• Kinetics and mechanism of nonequilibrium phase and structure formation in SHS • Mechanical activation in SHS
Institute of powder Metallurgy of the Byelorussian Academy of Sciences	• Mechanically activated SHS processes and synthesis of composite powders for protective coatings

Institute of technical acoustics	• Ultrasound effect
	• Composite materials obtained by centrifugal SHS casting
Byelorussian National Technical University	• Isostatic pressing of reactive powders for SHS of porous items
	• Steel coatings

Fig. (5.4.12): SHS in Byelorussia.

CHAPTER 6

Struggle for "Survival"

Abstract: The Chapter describes the attempts of the scientists in maintaining the SHS direction during the hard years after the USSR disintegration.

Our activity within MNTK was an outstanding stage in our life. It is always pleasant to know that your work is in demand. We were happy to advance and developed huge plans of our further activity.

But those plans could not come true. After the USSR disintegration in 1992 the system of MNTK also collapsed. We lost the government support. It was the main part of our maintenance. It was the hardest time for us. We did not have money to pay the salary. We lost a lot of our researchers: some of them went to work abroad (M.D. Nersesyan, E.A. Shtessel, A.S. Mukasyan, A.N. Filonenko, F.I. Rosenband, O.E.Kashireninov), the others started their own business.

After losing the state support, we lost our position of the head organization. MNTK disintegrated and each organization had to get out of the situation itself. Only God knows how we could survive – we did not have any experience in earning money. We were "a hothouse plant" – the state paid for everything and we did not have any problems. But we came to conscience after such a shock and began to work actively.

In the transition period we learned a lot and entered a new system of rapport, we continued advancing on the way of the scientific-and-technical progress.

Nowadays, ISMAN is still the largest and the most influential organization. Below you can see the structure of the SHS department of the Institute:

Laboratory headed by I.P. Borovinskaya
  Structural regulation of physical properties of SHS products
  Elemental SHS and SHS with magnesium reduction
  Nitride ceramics
  Immobilization SHS
  SHS technology
  Modeling of large-scaled SHS processes
  Capillary-porous structures
Laboratory headed by V.V. Grachev
  Development of filtration combustion theory
  Theory and experimental diagnostics of SHS processes in reactors
Laboratory headed by V.I. Ponomarev
  X-ray phase analysis
  Dynamic X-ray analysis
Laboratory headed by V.I. Yukhvid
  Liquid flame combustion of alumino-thermic SHS compositions
  SHS with centrifugal forces
  Technology of cast materials and items
Laboratory headed by A.S. Rogachev
  SHS in multilayered nano-sized films
  Mechanochemical activation of SHS
  Phase and structure formation
Laboratory headed by M.V. Kuznetsov
  SHS of oxides and oxide ceramics
  SHS in magnetic fields
Laboratory headed by T.I. Ignateva
  Chemical characteristics of SHS products
  Chemical analysis
  Chemical processing of SHS products
Group headed by A.F. Belikova

Physico-chemical analysis of SHS products
Laboratory headed by V.I. Ratnikov
Development of specialized SHS equipment
Standardization of SHS products
Laboratory headed by A.M. Stolin
SHS extrusion
Electric-spark alloying
Laboratory headed by V.A. Shcherbakov
SHS compaction
SHS welding

The most active are three laboratories headed by I.P. Borovinskaya, A.S. Rogachev, V.I. Yukhvid. (Fig. **6.1.**)

      I.P. Borovinskaya   *A.S. Rogachev*   *V.I. Yukhvid*

1. Laboratory of SHS Problems; Head of the Laboratory – Prof. I.P. Borovinskaya.

2. Laboratory of Microheterogeneous Process Dynamics; Head of the Laboratory – Prof. A.S. Rogachev.

3. Laboratory of Liquid-phase SHS Processes and Cast Materials; Head of the Laboratory – Prof. V.I. Yukhvid

Fig. (6.1): The most active SHS laboratories in ISMAN.

Inna Borovinskaya has already been working at SHS for 40 years. She took part in the research leading to the scientific discovery. Since then she has been really attached to SHS processes and passed the whole way from the scientific discovery to the production organization. She made a fundamental contribution to the theory of filtration combustion, developed the methods of chemical synthesis at the SHS mode, synthesized various nitrides and nitride ceramics herself, suggested and actively participated in the development of the SHS technology, fulfilled custom-made works, and introduced her elaboration into industry.

Inna Borovinskaya started working as a junior researcher then she became a senior researcher and the head of a small group. The group was growing fast and soon was reorganized to a laboratory. After defending her PhD thesis and preparing some young scientists, she became a professor. But there is one feature which prevented her from following the success of her life-work. It's her wonderful attitude towards her colleagues. She attracted young colleagues to the development of her ideas and if she saw that they had acquired a taste to independent work, she raised a question of organizing new laboratories for them.

Many of our famous scientists started their work at the laboratory of I.P. Borovinskaya. Among them are Alexander Rogachev, Alexander Mukasyan and others.

From the very beginning of his scientific career A.S. Rogachev has been interested in the work connected with experimental devices and obtaining and processing complicated information to be deciphered non-trivially. This work corresponds to his nature – he is a prudent, easy-tempered man. Being a creative person, he has always been looking for new solutions. When Inna told me about A.S. Rogachev (I had not known him well before), I decided to pass my laboratory to him. On the one hand I thought that a director could not have his own laboratory because he did not have enough time. Besides, my help to the members of the laboratory could be treated as a misuse of my official position. On the other hand I wanted to be closer to A.S. because we appeared to be like-minded people **[51]**.

Vladimir Yukhvid was a probationer from Kuibyshev Polytechnic Institute and worked in the group of E.I. Maksimov – an outstanding specialist in combustion. Vladimir investigated the combustion of thermite systems and liquid-phase mechanism of their burning. The aim of his work was to study the role of melt filtration from the combustion zone to the initial reagents. I was informed about his work. At that time we were keen on SHS and thought a lot how to develop our work. I had an idea to use the complicated thermite compositions, i.e. to synthesize cast refractory compounds by aluminothermic SHS processes. We discussed this idea with E.A. Maksimov and decided to offer the theme to Vladimir Yukhvid. He agreed. He started working enthusiastically. It was the first broadening of chemical syntheses in SHS **[49,50]**.

These three researchers have been always active. Some results of their work are shown in Figs. **6.2–6.9**. They include structural regulation of SHS product properties, optimum terms of SHS powder synthesis (it allowed us to raise the question about organization of the pilot plant of SHS powder metallurgy), SHS processes in multilayer nano-sized films ("non-powder" SHS), and application of SHS in the technology of heat-resistant alloys (for aviation) **[46, 48, 52, 56]**.

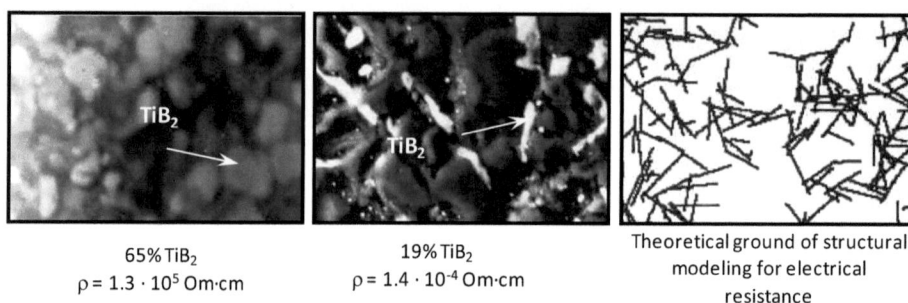

65% TiB$_2$
$\rho = 1.3 \cdot 10^5$ Om·cm

19% TiB$_2$
$\rho = 1.4 \cdot 10^{-4}$ Om·cm

Theoretical ground of structural modeling for electrical resistance

Fig. (6.2): Microstructure of TiB$_2$+AlN ceramics (I.P. Borovinskaya, E.A. Chemagina, A.Yu. Dovzhenko).

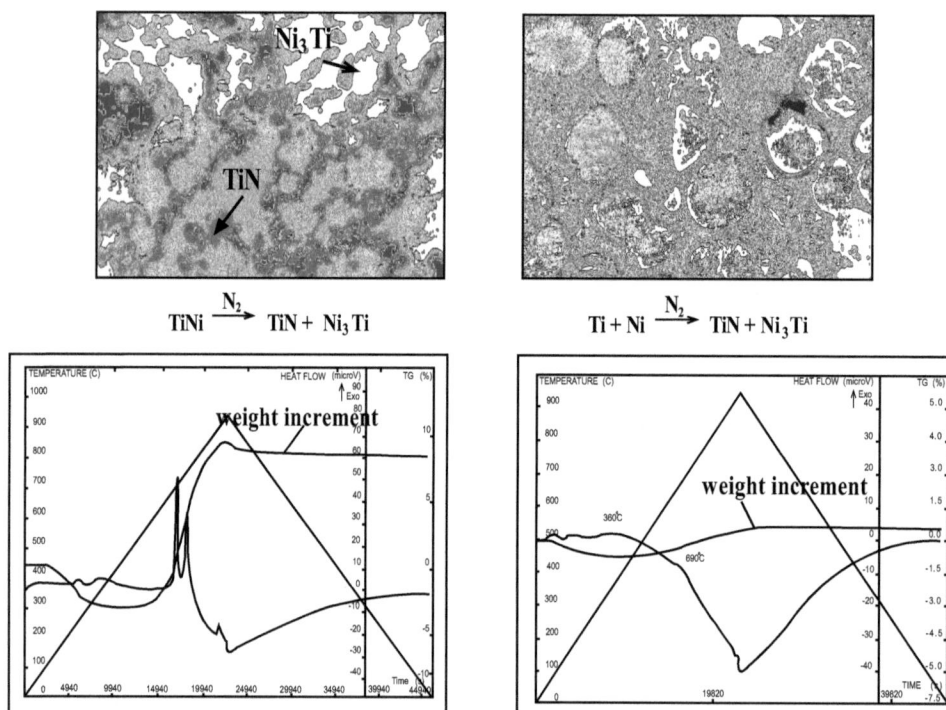

$$\text{TiNi} \xrightarrow{N_2} \text{TiN} + \text{Ni}_3\text{Ti}$$

$$\text{Ti} + \text{Ni} \xrightarrow{N_2} \text{TiN} + \text{Ni}_3\text{Ti}$$

Fig. (6.3): On structural regulation of corrosion resistance of TiN + Ni$_3$Ti ceramics. Regulation of SHS-product structure (I.P. Borovinskaya, E.A. Chemagina).

Quasi-homogeneous structure

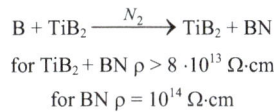

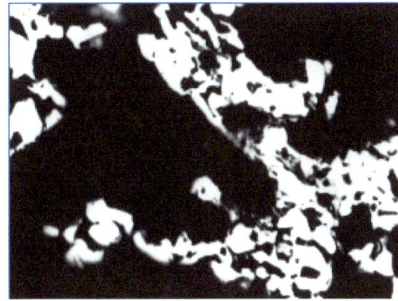

$$B + TiB_2 \xrightarrow{N_2} TiB_2 + BN$$

for $TiB_2 + BN$ $\rho > 8 \cdot 10^{13}$ $\Omega \cdot cm$

for BN $\rho = 10^{14}$ $\Omega \cdot cm$

Skeleton structure

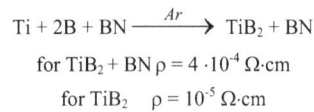

$$Ti + 2B + BN \xrightarrow{Ar} TiB_2 + BN$$

for $TiB_2 + BN$ $\rho = 4 \cdot 10^{-4}$ $\Omega \cdot cm$

for TiB_2 $\rho = 10^{-5}$ $\Omega \cdot cm$

Fig. (6.4): Structures of $TiN + Ni_3Ti$ double-phase product (I.P. Borovinskaya, V.A. Bunin).

Composition Trade mark / Parameters	AlN		Si₃N₄		TiC
	1E	**2 E**	**α-phase**	**β-phase**	**KT-1P**
Powders					
Chem. compos. % mass:					
C_{bound}	<0.02	<0.02	<0.1	<0.1	>19.5
C_{free}	-	-	-	-	<0.5
N_2	>33.5	>33.0	>38.8	>39.0	-
O_2	<0.5	<1.0	<1.2	<0.7	<0.5
Si_{free}	<0.07	<0.07	<0.2	<0.5	-
Fe	<0.1	<0.1	<0.06	<0.4	
Phase composition			>95% (α-phase)	>95% (β-phase)	
Specific surface, cm²/g	0.3–0.5	4–6	>6	>1.5	
Particle size, μm	<45	<0.1	Conglome-rates	Conglome-rates	(1000/630–40/0)
Products of SHS material processing					
Application	Substrates for electronics		Items of structural ceramic		Abrasive materials
Main parameters	Density 3.24–3.25 g/cm³ Hμ 13.2 GPa, Heat conductivity 100–130 Wt/(m·K)		Density 3.3–3.4 g/cm³ HRC 30 94–96 Fracture strength 7–9 MPa·m^(1/2) Bending strength 600–800 MPa		Abrasive ability (relative to steel) 62 mg Surface roughness (relative to brass) 0.20 μm

Fig. (6.5): Synthesis from elements. Some characteristics of SHS powders (I.P. Borovinskaya, V.V. Zakorzhevsky, V.I. Ratnikov, V.K. Prokudina).

Parameters

Composition Mark / Parameters	SHS-M		
	WC	**TiC**	**TiB2**
Powders			
Composition, % mass.			
C_{total}	6.1–6.2	>19.5	<0.1
C_{free}	<0.05	<0.4	
B	-		>30.6
O_2	<0.09	<1.0	<0.4
Si_{free}			B_2O_3 <0.2
Fe		<0.04	<0.35

Mg total	<0.02 (total)	<0.5	<0.2
Mg Acid solution		-	
Spesific surface, m²/g		<7.9	2
Particles size, μm	<1.0	<1.0	-10>70% -25>90%
Products of SHS material processing			
Application	Hard alloys of VK6 type	Tungsten-free STIM alloys	Tungsten-free alloys STIM-1 Evaporative elements
Main operation parameters	Density – 14.9 g/cm³ HRA – 91 Bending strength – 1700 MPa Stability coefficient 1.4	Density – 4.5–5.9 g/cm³ HRA – 90–93 Bending strength – up to 1100 MPa	Density – 4.7–4.9 g/cm³ HRA – 92–93 Bending strength – 400–800 MPa

Fig. (6.6): Magnesium-reduced synthesis (SHS-M). Some characteristics of SHS powders (I.P. Borovinskaya, V.I. Vershinnikov, T.I. Ignatieva).

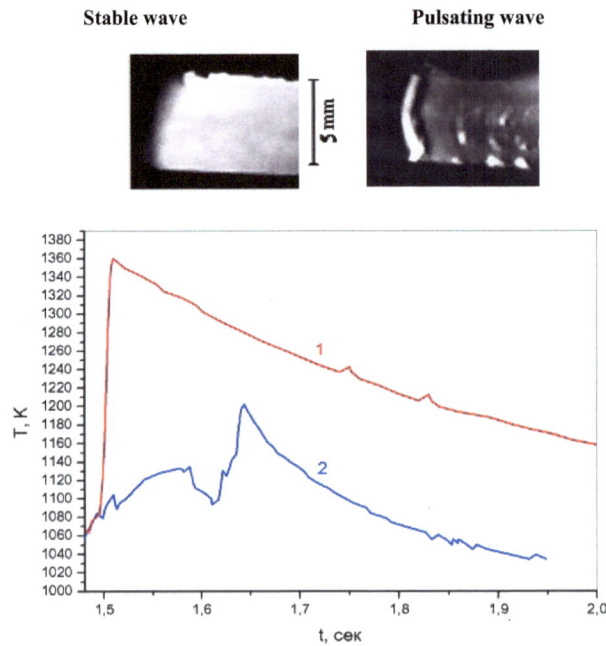

The temperature profile of stable (1) and pulsating (2) waves

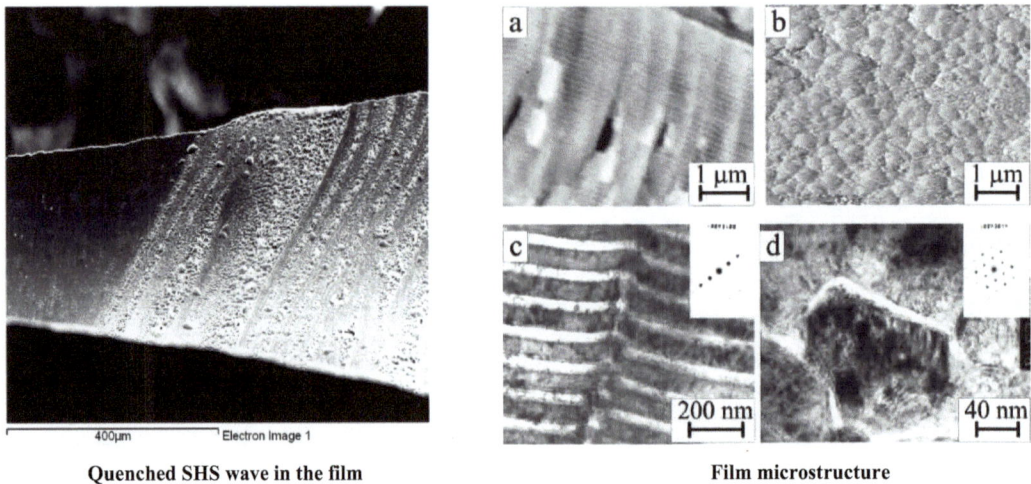

Quenched SHS wave in the film   Film microstructure

Fig. (6.7): Gasless combustion of multilayer nanofilms: Ti/Al composition (A.S. Rogachev, H.E. Grigoryan, J.-K. Gashon *et al.*).

I. CAST HEAT-RESISTANT ALLOYS

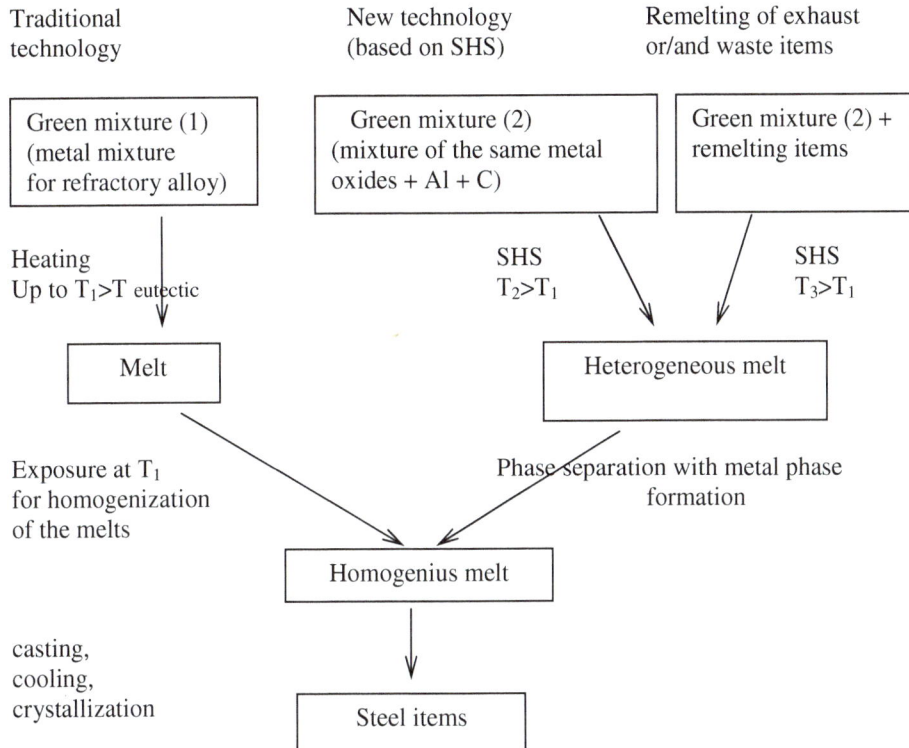

Efficiency: application of cheaper raw materials, increase in production output.

II. CAST RODS AND MOLDS

Application of SHS products (without changing the available technology)

Item	Efficiency
Molds SHS "Ruby" (Al_2O_3-Cr_2O_3) instead of Al_2O_3 Laboratory headed by Prof. V.I. Yukhvid	• Increase in heat conductivity • Decrease in grain size • Improvement of surface quality
Rods SHS "Ruby" (Al_2O_3-Cr_2O_3) + SiAlON (instead of Al_2O_3) Laboratories headed by Prof. V.I. Yukhvid and Prof. I.P. Borovinskaya	Absence of fluorine-containing solvent of Al_2O_3

Fig. (6.8): SHS in production of heat–resistant alloys (V.I. Yukvid, Yu.S. Eliseev, A.G. Merzhanov, V.N. Sanin, V.A. Poklad, O.G. Ospennikova, V.V. Deev).

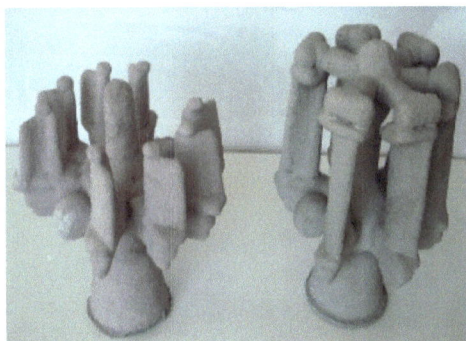

molds

rods

turbine blades

Fig. (6.9): SHS in production of heat-resistant alloys: items.

A huge work was done by T.P. Ivleva in 3-D modeling of the waves of solid-flame combustion. We worked together. We found a lot of unknown wave structures and its propagation modes. We presented them in a special album because we think that they are of the artistic interest. In Figs. **6.10, 6.11** you can see various types of non-stationary temperature fields and the conversion degrees in the wave. Nowadays we are working at the scientific generalization of these unique results [57].

I have always realized that a problem solution must be guided by the analysis of appropriate experimental-theoretical models. I even developed the concept of 4 levels of adequacy:

Level 1 implies qualitative conformity between experimental and theoretical data. For example, the fact that the auto-oscillating front or spin waves exist is enough to say about the 1-st level of adequacy.

Level 2 – qualitative conformity of combustion regulations. E.g. the combustion rate dependence of particle size: it is decreasing both in experiments and in the theory.

Level 3 – quantitative description of the process regularities. In this case functional dependences of characteristics on parameters should coincide. E.g. if according to the experimental results and in the theory the dependence of the combustion rate on the particle size corresponds to the parabolic law U ~ r-1 we can assert about the adequacy of the third level. If all the dependences on the parameters coincide, we deal with the complete realization of the 3-rd level adequacy, if not – partial.

Level 4 – complete coincidence of absolute values of any process characteristics. Such a level was achieved in our work on thermal explosion of the substance "Dina" where we could create the experimental model strictly corresponding to the model of thermal explosion in N.N. Semenov's theory. The temperature values calculated theoretically and those obtained experimentally differ by 1° maximum.

In view of the aforesaid I'd like to mention the work of V.V. Grachev. He graduated from Kuybyshev Polytechnic Institute and specialized in combustion theory. He began his work and cooperated with E.N. Rumanov in the development of the theory of combustion and explosion processes. But also he became interested in experimental study of the processes in technological SHS reactors. This work could belong to the 3-rd level of adequacy.

Some results obtained by Dr. V.V. Grachev are shown in Fig. **6.12 [58,59]**. He developed the mathematical model of a reactor used for synthesizing nitrides under nitrogen pressure. In such a reactor nitrogen is above the initial mixture layer. The combustion wave propagates along the mixture, and the gas is filtrated to the front in the transverse direction. The important parameters in this model are dimensionless values A and D. A is the dimensionless coefficient of heat exchange between the green mixture and the gas, D is the ratio of the gas mass in the reactor to that necessary for complete conversion of the solid reagent. V.V.

Grachev marks out three zones in the (A, D) plane. In zone 1 the gas pressure grows due to the gas temperature increase during the reaction. In zone 3 the gas pressure

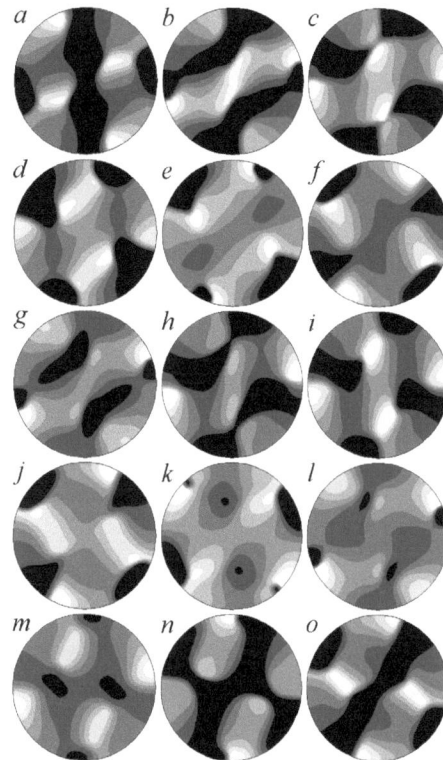

Fig. 6.10: 3-D modeling of combustion. Succession of cylindrical sample's sections at spin wave propagation. Lighter areas correspond to higher temperatures (T.P. Ivleva, A.G. Merzhanov).

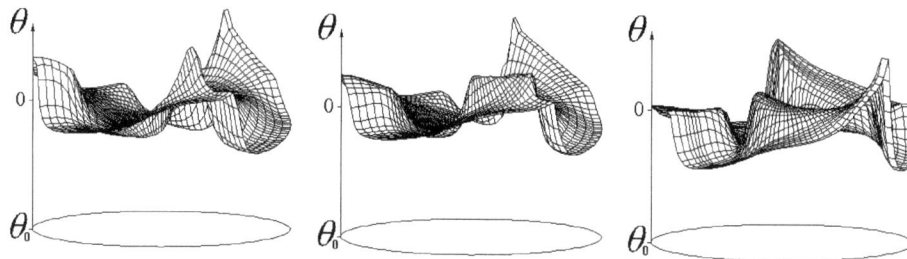

Fig. (6.11): 3-D modeling of combustion. Evolution of isothermic surface at chaotic combustion wave propagation along a cylindrical sample (T.P. Ivleva, A.G. Merzhanov).

drops due to nitrogen absorption. And in zone 2 the gas pressure is approximately the same since the opposite effects of the gas temperature and absorption compensate each other (in the figure it is the zone close to the curve). Zone 2 is recommended to be realized under technological terms because it allows obtaining a homogeneous product. V.V. Grachev carried out some experiments in silicon combustion in nitrogen. They confirmed his theoretical prediction. In these experiments the product homogeneity was characterized by a-Si3N4 content in the surface layer and in the center of the green mixture. Below you can see the measurement results:

-   In zone 1 α-Si_3N_4 content in the surface layer was 70 %, in the center a-phase was not found, the product consisted of b-Si_3N_4. The product phase composition was not homogeneous.

-   In zone 2 α-Si_3N_4 content in the surface layer and in the center ranged from 65 to 72 %, the product was rather homogeneous.

-       In zone 3 α- Si_3N_4 content in the surface layer and in the center ranged from 68 to 75 %, but a considerable quantity of unreacted silicon was obvious.

In 1997 we celebrated the 10-th jubilee of our Institute. We wanted to thank everybody who helped us. We made memorable medals for them by the SHS method. I made a report on the history of our Institute, its work and about everybody who helped us. When I mentioned somebody, I interrupted my speech and presented him the medal, and then continued the report. I'd like to mention the names of these helpers: F.I. Dubovitsky, A.E. Shilov, G.I. Marchuk, V.A. Kabanov, Yu.A. Osipiyan. All of them are outstanding scientists.

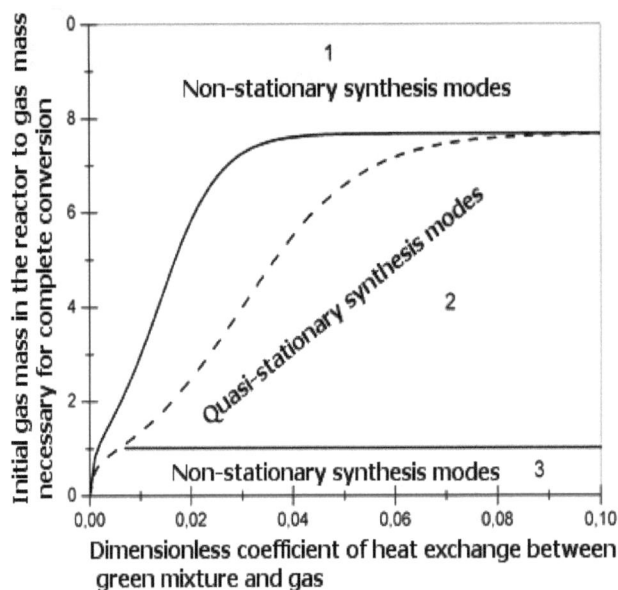

Fig. (6.12): Fragmentation of parameters' plane to the zones corresponding to various modes of nitride synthesis in the SHS reactor: 1 – non-stationary modes at continuous growth of gas pressure during combustion; 2 – quasi-stationary modes when maximum gas pressure in the reactor is achieved before the end of the process; 3 – non-stationary modes at continuous decrease of gas pressure during the combustion. The zone close to the dotted line corresponds to the optimum terms of the synthesis.

In the period of economic dislocation prevailing in the post-Soviet area it was not only ISMAN which could survive, there were some other organizations being on their feet during the Soviet time. They are ISMAN Branch in Tomsk, SHS Educational Center of MISiS-ISMAN, Engineering SHS Center of Kuibyshev Polytechnic Institute (in Russia), Kazakh Institute of Combustion, Armenian Institute of Chemical Physics. Georgian SHS was being developed. Belorussian SHS was formed. The SHS Center was organized in Barnaul.

And now I'd like to dwell on the development of SHS all over the world. I think it is the most important stage [30-41].

CHAPTER 7

SHS All Over The World

Abstract: The Author considers the international development of SHS and the results obtained in other countries – the USA, Japan, etc.

In 1980 SHS investigation started abroad – in Japan and the USA. What was our attitude to it? Of course, we were glad. We realized that SHS would be developed faster and we would get not only new results but innovative views and approaches.

I had a question – why so late? We had already published some articles, taken out several patents. Later I asked Dr. Holt, one of the leaders of SHS in the USA, about it. He answered without hesitation: "We knew about your remarkable investigation but we considered them rather exotic and useless. And only when you started the production we paid serious attention to your discovery" **[60, 61]**.

Some years later one of the members of the Japanese company "Chory" explained the situation almost in the same way. He said: "When I learned about the process, I studied the available literature and realized that the process was very interesting. I wanted to have it in Japan since we have the highest technological level and know very well what is acceptable for industry. I invited some professors from our Universities and scientists working for the companies and delivered a lecture on SHS. Their reaction was unanimous – the process is splendid but it can't be used in industry since it is non-equilibrium and uncontrollable therefore it won't provide good products".

I did not object to it. I realized that when working "all alone", without any competition with the West, we managed to do a lot and created a good start in many directions.

7.1. SHS IN USA

As I have written already we were impressed by the article of Joy Crider about our process **[5]**. It was written to induce American scientist to start investigating the process. But it appeared to be a good impulse for our statesmen to help us.

Crider was entrusted to write the article by J. McCauley who issued 15 papers about SHS during its initial stage in the USA. We consider him to be the Pioneer of SHS in the USA.

A decisive role in the SHS development belonged to John Kiser. His company "Kiser Research" was founded for establishing business contacts between American and Soviet scientists with the aim of SHS development in the United States on mutually beneficial terms. After getting acquainted with us and the SHS process, Kiser began to advance our achievements. He organized our lectures in various American Universities, arranged bilateral symposia. Thanks to his activity I got acquainted with many colleagues, visited different cities in the USA, and delivered a lot of lectures in SHS. It stimulated the interest of our American colleagues to SHS development.

When the American scientists began working in this field, they changed the name of the process. They called it Combustion Synthesis. I have some comments. I have always thought that it is not ethical to change the name given by the authors. But there is another consideration. At first we wanted to connect the name of our process with combustion and began to call it Solid Flame. We also thought to call the process Combustion Synthesis analogously to Explosion Welding. But we realized that the important role in the process belonged to structure formation and we wanted to reflect its essence more exactly and thought out a more accurate but longer name "self-propagating high-temperature synthesis".

The famous Lawrence Livermore Laboratory was entrusted to develop the problem. The Head of the program was Birch Holt. A little bit later Professor Zuhair Munir joined him.

Nowadays the USA is a country with prominent R&D in the SHS field. It is the second after Russia. I'd like to introduce you the American scientists who are the most active in this field.

Z.A. Munir is one of the most competent American scientists working in SHS. He started rather modestly. Some of his first results resembled those which were achieved in our country. But then Prof. Munir carried out some original works and showed the variety of his interests. Recently he took a great interest in electric field influence, i.e. additional energy introduction to the combustion zone (Fig. **7.1.1**). Prof. Munir is a popular and famous propagandist of SHS. There are always a lot of students and young scientists in his laboratory in the University of California. He has received numerous awards and honors. Professor Munir has published more than 420 papers and edited 8 proceeding volumes. He holds 13 US Patents. In 2003, the Institute for Scientific Information (ISI) listed him as a Highly Cited Author in Materials Science **[62, 63]**.

Bernard J. Matkowsky is a well-known scientist working in the field of applied mathematics used for solving combustion theory tasks. He was engaged in the study of cellular flames. Being interested in mathematical problems of combustion theory of SHS systems, B.J. Matkowsky invited the scientists from Chernogolovka – K.G. Shkadinsky, V.A. Volpert, A.P. Aldushin, who had an experience in theoretical and mathematical study of SHS processes, to work in the University of Evanston (USA) under his leadership. This work appeared to be very fruitful. In their joint investigations they considered models of combustion wave propagation with due regard to melting of one of the reagents; they described the peculiarities of liquid-flame combustion, developed the theory of filtration combustion, thoroughly studied smoldering as a mode of filtration combustion. This team published 32 papers by SHS **[64]**.

Prof. Mark Meyers has carried out research since 1972 in a broad range of areas within Materials Science including dynamic processing (explosive consolidation, synthesis, welding, shock- and shear induced reactions, and combustion synthesis), dynamic fracture and fragmentation, dynamic and shock response of materials.

He has made contributions in martensitic transformations, twinning, constitutive equations; he studied the influence of grain size on the strength of metals, ultrafine and nanocrystalline metals, and mechanical properties of biological materials. In the area of combustion synthesis he developed a technique of dynamic consolidation of TiC, TiB2, and ceramic composites using a high velocity forging press (Fig. **7.1.2**); investigated micromechanisms of reactions at the combustion front; successfully produced and characterized TiC-NiTi cermets by combustion synthesis. Professor Marc Meyers is Co-founder of the Center for Explosives Technology Research and EXPLOMET conferences. He is the author or co-author of 290 research papers and three books. Professor Marc Meyers is the Leader in the field of dynamic SHS densification **[65]**.

The method allows:

- Obtaining high pure ceramic and cermet materials with microstructural characteristics which can't be achieved during the sintering method.

- Modeling a change in an item's dimensions analytically using the homogeneous approach (it makes it possible to obtain items of a preset shape).

- Determining material characteristics by comparing loading-deformation experimental curves with calculated ones.

- Using the method of modeling by final elements for predicting the density distribution inside a porous sample during mechanical loadings.

Z.A. Munir

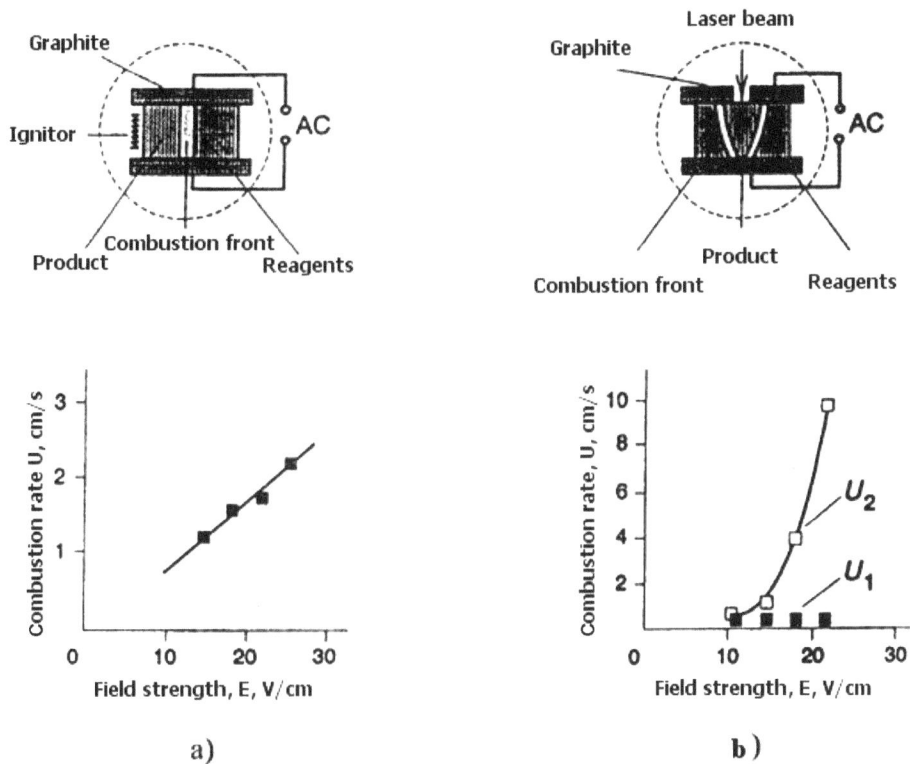

Fig. (7.1.1): SHS process stimulated by additional electric power (Field Assisted Combustion Synthesis). Electric field can be applied perpendicularly to the combustion direction (a) or parallel with it (b).

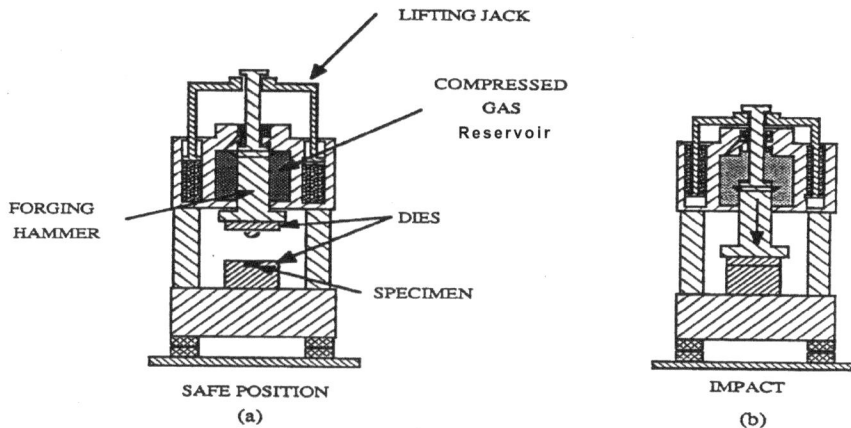

Fig. (7.1.2): Forced compaction by M.Meyers.

Professor Puszynski has been one of the pioneers of self-propagating high-temperature synthesis (SHS) in the United States of America. His work in that field began in 1982 when he joined the State University of New York at Buffalo, NY. During his stay at this university he jointly with Professor Hlavacek designed and built SHS reactors for commercial production of aluminum nitride at low nitrogen pressures. The main result of their activity was elaboration of the mathematical model which described exothermic reactions in "solid-solid" and "solid-gas" systems. They described a complicated structure of the combustion front in non-one-dimensional situations. He also studied silicon and aluminum nitride synthesis at combustion of silicon and aluminum in nitrogen. In 1991, Dr. Puszynski accepted the faculty position at the South Dakota School of Mines and Technology in Rapid City where he continued the work in the area of SHS and other combustion related areas. At the South Dakota School of Mines and Technology he focused his research on the combustion synthesis of complex solid solutions and composite materials based on Al-Si-Ti-N-C-O system.

Professor Puszynski's major contributions into the SHS field include:

1. Nitridation of aluminum, silicon, boron, and transition metal nitrides;

2. Formation of α- and β-sialons, alon, tialon, and $Ti_xAl_yC_z$ solid solutions;

3. Densification of combustion synthesized TiC-TiB_2 composites;

4. Simultaneous combustion synthesis and densification of ceramic and intermetallic composites in the field of uniaxial and centrifugal forces;

5. Formation of nanopowders and nanocomposites in a self-sustaining regime;

6. Investigation of combustion front propagation in systems consisting of nanoreactants with and without dispersed carbon nanotubes;

7. Mathematical modeling of self-sustaining processes in both gasless and gas-solid systems.

During his professional career Professor Puszynski has published over one hundred thirty peer-reviewed papers and presented more than two hundred papers at the national or international conferences and seminars. Some results are presented in Figs. **7.1.3** and **7.1.4 [66,67]**.

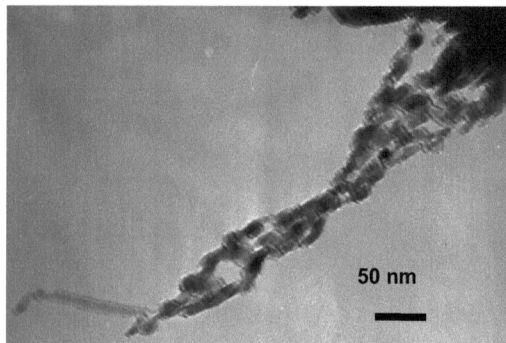

Fig. (7.1.3): TEM microscopy: carbon single nano-tubes and nickel aluminide nano-sized grains on the bunches of these nano-tubes can "survive" in the combustion wave (by J. Puszynski).

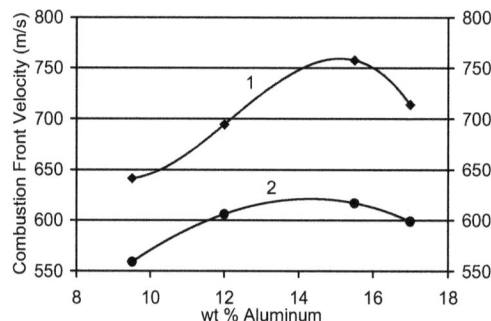

Fig. (7.1.4): Combustion front velocity in Al-Bi_2O_3 (1 – initial aluminum had a protective coating of hydrophobic oleic acid (by J. Puszynski).

Prof. John Moore pays great attention to the development of advanced materials and their commercial application.

The most significant directions in SHS research for him are:

*   Synthesis of porous materials for bone replacement applications

*   Joining of dissimilar materials and materials that are difficult to weld or join.

- Coupling of SHS with densification/ consolidation processes in the production of dense composite materials for many commercial applications, e.g., wear resistant ceramic matrix materials, high performance metal matrix materials, armor plating, PVD targets.

- Study of the effect of microgravity on SHS to facilitate 'in-space fabrication and repair' (ISFR) and "in-space resource utilization" (ISRU), sterilization of planetary minerals

- Synthesis of ceramic nuclear fuels

- Development of auto-ignition combustion synthesis (AICS) for the production of nanoscale oxide and nitride powders

- Synthesis of dense and porous shape memory/super elastic materials based on NiTi and composites of NiTi with TiC and TiB$_2$.

He is one of the leaders in creating and applying advanced materials for commercial properties **[68]**.

Prof. Thadhani is a famous American scientist, a specialist in mechanics of deformation processes in solid media. He developed a new direction connected with a shock effect on SHS green mixtures. He provided the distinction between chemical reactions occurring during a high-pressure shock state (termed "shock-induced") and those occurring after unloading of the shock pressure to ambient conditions (termed "shock-assisted" or "shock-activated"). He has applied the shock-induced reactions for synthesizing high-pressure phases and non-equilibrium compounds, and for studying a new class of advanced energetic/reactive materials **[69]**. The shock-assisted reactions have been applied for processing nanocrystalline intermetallics and fine-grained ceramics. Using shock-activated SHS he has solved the problem of nanocomposite magnets.

He is the author of more than 150 articles and an active organizer of multiple workshops and conferences.

James McCauley   *J.B. Holt*   *B. Matkowsky*

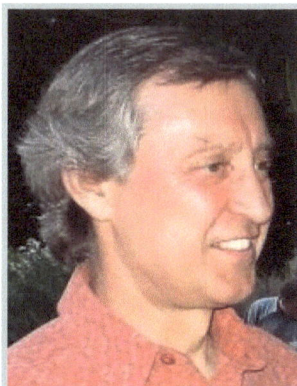  

J. Moor   *J. Puszynski*   *N. Thadhani*

Fig. (7.1.5): American Pioneers in SHS.

Prof. Dan Luss investigates chemical technologies in the University of Houston. His main interest is synthesis of oxides and formation of electric and magnetic fields during SHS.

Along with Karen Martirosyan he worked out an SHS process which was called carbon burning of complex oxides. This method was shown to have some advantages in comparison with the conventional SHS. For instance, it allows obtaining powders of a smaller particle size.

The group headed by Prof. Luss studied formation of electric fields during SHS of powder mixtures and separate particles. They also measured magnetic fields and developed a theoretical model predicting the electric field formation. Prof. Dan Luss got a lot of professional awards for his scientific activity [70,71].

I'd like to introduce you Kathrin Logan. She has been engaged in SHS processes for a long time. Unlike other American scientists she does not dissipate her energies between different themes; she does not concern herself with all the problems coming to mind; she is devoted to only one "gentleman" – titanium diboride and its relative – cermet product $TiB_2 + Al_2O_3$. She has substituted the breadth of interest for the depth of experience. Kathryn Logan has determined the SHS reactions which form these products. She used the methods of thermodynamic calculations, studied kinetics of the reagent interactions as well as mechanism of combustion and structure formation. She carried out a lot of syntheses and defined the optimum conditions, developed the technology of the powder production. She created a pilot production line, realized marketing with delivery of powder pilot lots and feed-back about the obtained results, determined the technical and economical efficiency and established commercial production. One of the most favorite expressions of Kathryn is "it is so pleasant to speak about what has been done but not about what should be done!"

Particularly, I'd like to tell you about Prof. Alexander Mukasyan. He studied SHS in Chernogolovka. He began his activity in the laboratory headed by Inna Borovinskaya. He was engaged in synthesis of ceramics items. From the very beginning he showed his excellent features, such as inquisitiveness, quick-wittedness and capacity for work. He took an active part in the development of an excellent material, so called black ceramics.

When Inna Borovinskaya realized that he is ready for independent work, she asked me to create a new laboratory for A. Mukasyan but I did not have an opportunity at that time and invited him to my laboratory. Some time later we organized a laboratory for him and he headed it in 1994–1996.

But he has displayed his talent in full in the USA where he lives and works now. He is engaged in R&D activity at the University of Notre Dame, IN, USA, in the group headed by Prof. Arvind Varma, an outstanding scientist in chemical engineering. Fundamental mechanisms of combustion wave propagation, kinetics of rapid heterogeneous reactions, SHS in microgravity and solution combustion synthesis are the main scientific directions for him during the last ten years. His current scientific interest is primarily related to nanotechnology and novel alternative energy sources including fuel cells, hydrogen production and storage. He has published more than 125 scientific papers in various international journals. Now he is a professor of the department of chemical and biomolecular engineering at the University and considered to be one of the leading scientists in the SHS field [72,76].

I'd like to give an example. When developing the ideas of Prof. S.K. Patil (India) in solution combustion synthesis of nano-materials, Prof. A.S. Mukasyan and his colleagues created an original method for obtaining nano-powders – combustion synthesis of impregnated active substrates [77].

The essence of the method is rather simple (Fig. **7.1.6**). A porous structure which can burn in the air yielding gaseous products (e.g. cellulose or carbon nano-tubes) is impregnated by a reactive solution of a required composition (Stage 1). Then water excess is removed from the obtained heterogeneous medium (Stage 2). And within Stage 3 the reaction is locally initiated and the SHS wave propagates along the mixture. As a result a single solid-phase product is obtained, it is an oxide nano-powder of the composition corresponding to the calculated one. The support has several functions: 1. It appears to be an additional heat source and allows carrying out the synthesis in low-exothermic systems; 2. It converts the combustion mode from heat explosion to stationary reaction wave propagation and allows controlling the synthesis and achieving 100 % output; 3. It enables starting continuous synthesis. This method resulted in obtaining new catalysts which demonstrated unique properties.

But also I'd like to mention that there are many small groups all over the USA which are engaged in SHS. Among them are the teams headed by Prof. T. Weihs at the University of John Hopkins (combustion of nano-sized multi-layer films), Prof. K. Brezinsky at Illinois University (nitride synthesis in flow reactors), Prof. S. Son at Purdue University (synthesis of high-porous foam materials), Prof. S. Bhaduri from Clemson University (obtaining of nanopowders by solution combustion), etc.

In the conclusion I'd like to underline that Russian specialists rate highly the results achieved by their American colleagues. During the work we always maintained friendly relations and organized a lot of common scientific events (Fig.**7.1.5**).

A.Mukasyan

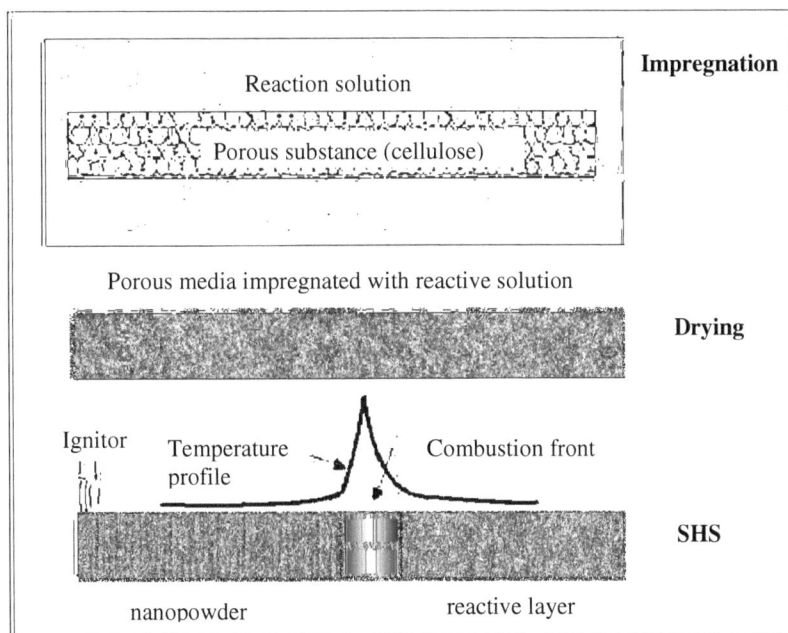

Fig. (7.1.6): Synthesis of nano-materials by combustion of impregnated active supports.

7.2. SHS IN JAPAN

Our Japanese colleagues started their SHS investigations in another way. "Chory" Company was going to buy a license from us for producing SHS powders. They concluded an option agreement with "Licenzintorg" on synthesizing SHS powders of 5 types. We carried out the work excellently. The powders were highly evaluated by Japanese experts. Then "Chory" made a business proposition to "Licenzintorg". But an unexpected event occurred. After getting the money for the option agreement, the Academy of Sciences did not allow "Licenzintorg" to sell the licenze. Then the Japanese specialists who knew about thatproblem decided to start their own investigation of the SHS process. They organized a team and invited the venerable scientist – Prof. M. Koizumi – to be at the head. Such well-known researchers as Miyamoto and Odawara joined the team (Fig. **7.2.2**).

Prof. Koizumi paid much attention to technological elaborations. At first they worked in two centers – the University of Osaka (Koizumi and Miyamoto) and the University of Science and Technology in Tokyo (Odawara). Then they organized the work in the University of Ryukoku.

I'd like to mention the most prominent technological results of our Japanese colleagues:

1. A technology was developed for producing long pipes with an internal wear-resistant coating (O. Odawara). Such a pipe was obtained when a green mixture containing iron-aluminum thermite was burnt up in an ordinary steel pipe of 8 m in length and 250 cm in radius. Two-phase melt was formed during the process (Fe + Al_2O_3). Due to the centrifugal force the melt separated into layers and moved to the pipe walls. The external layer of the melt (Fe) provided strong adhesion with the initial pipe, and the internal layer (Al_2O_3) was a wear-resistant coating. When I was introduced to Prof. Odawara I called him "Mister Long Pipe". He liked the joke **[78,79]**.

2. A new variation of forced SHS compaction was worked out. It is a uniform compression of a burned-out sample in a shell (SHS HIP, Miyamoto). In this case a sample is enclosed into a deformable gas-proof shell and placed into a constant high pressure vessel (we call it "a gasostat"). After the combustion process the green mixture is compressed to almost non-porous state but it keeps its shape with a slight decrease in its size due to the pore space disappearance **[80]**.

3. A technology of functionally graded materials (Koizumi, Oyanagi). At first a multilayer green mixture is composed. There is a special device for preparing such a mixture automatically. Each layer differs slightly from the previous one in its composition. Then the mixture is ignited and the combustion process yields a functionally graded material (Fig. **7.2.1**) **[81]**.

    We used another principle for this aim (A.N. Pityulin et.al). It is more original but less universal. One of the examples is TiC+Ti plate production. A two-layer green mixture is prepared; one layer consisting of titanium and soot, the other – of nickel. During the combustion process nickel melts and saturates the burnt mixture of titanium and soot, i.e. the formed titanium carbide. A graded composition is formed within the saturation. It is possible to arrest the process at any stage of saturation by applying pressure and densifying the plate **[82]**.

4. Growth of single crystals of non-oxygen refractory compounds obtained from SHS powders (Otani). Single crystals, grown from ordinary furnace-made powders, were of low quality and those obtained from highly pure SHS powders were of high quality.

5. Obtaining of solid hydrogen accumulators (intermetallic hydrides) by burning intermetallic compounds in hydrogen.

Also I'd like to mention the work of Dr. Makino who is engaged in the development of the theory of SHS system combustion. His results in propagation of non-adiabatic flames make an important contribution to combustion science **[83]**.

Nowadays Japanese scientists work actively in the SHS field. Some new scientists have joined us and their works are of interest too (e.g. synthesis of complex compounds such as $Ba_2Si_5N_8:Eu^{2+}$ (diode emitter) and $CaAlSiN_3:Eu^{2+}$ (luminophor).

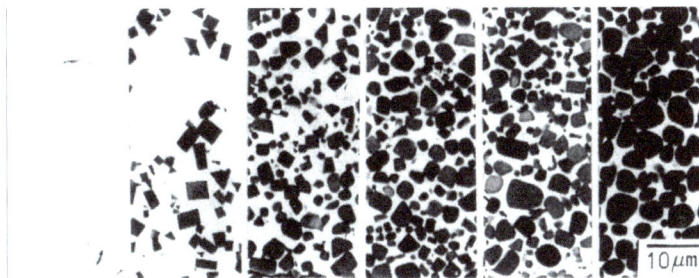

Fig. (7.2.1): Graded phase distribution in TiC-Ni. Left – 100 % nickel content, right – 30 % nickel content. Dark grains – titanium carbide. (Yo. Miyamoto with co-autors).

    M. Koizumi   **O. Odawara**   **Yo. Miamoto**

Fig. (7.2.2): Japanese researchers in the SHS-field.

7.3. SHS IN CHINA

After getting acquainted with our information on SHS, some Chinese scientists tried to study this new direction.

But serious assimilation of the process started after a visit of the Soviet delegation (I.P. Borovinskaya, Yu.M. Maksimov, S.S. Mamyan, N.S. Makhonin) to China for a meeting with young Chinese scientists. It was organized by All-China Metallurgical Association (Prof. Ma). Each of the members of our delegation delivered several lectures to our Chinese colleagues.

Two SHS centers were founded in China. One of them was in the University of Science and Technology in Beijing, the other – in Wuhan. The latter was headed by Prof Yuan. We had close collaboration with Prof. Yuan – we organized joint work in direct SHS production of hard materials and created a research center for it. According to this program one of our specialists Dr Pityulin worked in China during a year [84, 85].

In 1995 we held the III International SHS symposium in China; it appeared to be a great stimulus for SHS development in this country, Japan and Korea. As for me I had friendly relations with Prof. Yuan and we highly valued these relations.

Prof Yuan had a favorite follower – a young scientist Fu. When he was going to defend his PhD thesis, Professor asked me to be the Chairman of the Scientific Council. I agreed. During the presentation the members of the Council were very inactive. They did not ask any questions. I felt embarrassing. I managed to force the members to ask questions and then I made them appreciate the thesis. So the presentation was very interesting and unusual for the University. Many members of the Council thanked me and invited to participate in such events.

After the death of Prof. Yuan (it was a great loss for the Chinese SHS) Prof. Ge working in the University of Science and Technology in Beijing became the leading specialist in SHS **[86]** (Fig. **7.3.5**).

His colleagues carry out the work in:

- SHS of ceramic materials;
- Solution combustion yielding nano-sized oxides;
- Frontal polymerization (the processes which is analogous to SHS).

In (Figs. **7.3.1** and **7.3.2**) you can see some results of his research of SHS ceramics and solution combustion. His work in frontal polymerization was highly appreciated by his colleagues all over the world. I'd like to quote the words of a reviewer from Chemistry-A European Journal: "This is a very, very, very nice paper".

I'd like to mention another SHS center in China – Harbin Institute of Technology. We helped them with equipment, acquainted with the SHS backgrounds and organized their young specialists' probation at ISMAN. The leader of the work Prof Du and his young and active colleagues created an excellent base further SHS development. During his visit to China the Vice-President of the Russian Academy of Sciences academician O.M. Nefedov was impressed by their work. In many Chinese universities which he visited, he heard about the results in SHS. Once he asked: "Do the scientists in all the Chinese Universities and Institutes study SHS?" And he heard in response "almost".

We had a lot of contacts with various scientific organizations in China. But not everything went on smoothly. There were some failures. For example, by the initiative of the Chinese Metallurgical Association we founded a joint venture for SHS powder production. We met our engagements and our colleagues left for Beijing to organize the production line. Everything was O.K. But at the beginning of the commercial stage when the Chinese partners concluded that they could do everything themselves, the products appeared to be of a low quality and could not be sold in the market. The joint venture failed. Then a new company was organized, it used the same equipment, the same raw materials, the same technological specifications, but the products obtained were much better and could be sold with no problems for some reason.

Fig. (7.3.1): Submicron particles of β-SiC (Prof. C.C. Ge).

Material	Fuel	Particle size	Application
Gd_2BaCuO_5	CA/EDTA	50-100 nm	high-Tc superconductors materials
TiO_2	GLY	50 nm	photocatalytic materials
CeO_2-TiO_2	CA	15-25 nm	photocatalytic materials
RE-CeO_2	CA/GLY	13-25 nm	oxide-ion conducting electrolyte
$Ba_3(Ca_{1.18}Nb_{1.82})O_{9-\delta}$	CA	<50 nm	solid proton conducting electrolyte
RE-$BaCeO_3$	CA	30-60 nm	solid proton conducting electrolyte
$BaTiO_3$	CA	<50 nm	electro-ceramic material

Fig. (7.3.2): Oxide materials obtained in solution combustion mode (C.C. Ge with co-others).

We had an incident in Harbin. At first the work was considered to be a joint one but then it appeared to be custom-made. We helped our Chinese partners to make an excellent basis, taught their young specialists in our Institute in Chernogolovka. But when they were leaving us they took a great number of preprints of our papers in Russian.

But they did not manage to organize their work without any conflicts. At first the researchers in Harbin SHS Center repeated our work; frankly speaking, they translated our papers into English and published them as their results. We noticed it and expressed out indignation. But soon they started working by their own, and our conflict was over.

But these "Chinese misunderstandings" did not shake our respect to great Chinese people.

I'd like to mention two peculiarities of Chinese SHS:

The first one is connected with high rates of self-propagation of SHS investigation all over the country and rapid growth of the number of works fulfilled and researchers involved. The second one consists in their great wish to get the final product and create the production line. Especially impressive is their fast advance from basic research to industrial implementation in production of large pipes with inner wear-resistant coatings. The idea suggested by Prof. Odawara was developed and then realized for large pipes in China. The production was successfully organized due to the wholesome competition between Prof. Sheng Yin (Beijing) and Shu Ge Zhang (Nankin). The photos of the pipes (Fig. **7.3.3**) are very impressive [87,88].

Fig. (7.3.3): Industrial production of pipes and pipelines for transportation of abrasive materials in China (Sh-G. Zhang).

A very promising scientist is Prof. Fu. Fig. **7.3.4** demonstrates the data of titanium diboride production developed by him.

To date China is the most suitable country for large-scale SHS production due to its rich ore and raw material sources and wise policy of the government.

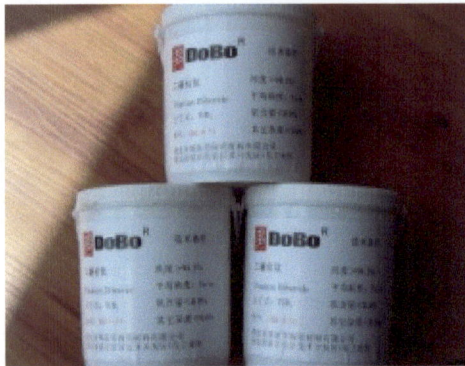

Fig. (7.3.4): Commercial SHS products made by Hubei DoBo Advanced Ceramics Co., Ltd.: titanium diboride powders (40 000 kg/year), evaporative boats (200 000 un/year).

R.Z. Yuan *Sh.-Ge. Zhang* *C.C. Ge*

Fig. (7.3.5): Chinese SHS specialists.

7.4. SHS IN OTHER COUNTRIES

Nowadays SHS investigations are carried out in more than 50 countries all over the world. The most active countries are the USA, Japan, and China. But I have already described their achievements. I would like only to add that in these countries there are even National SHS Associations.

As to other countries, there are only separate groups or a single activist.

First of all, I'd like to mention a Ceramic Group in Cracow (Poland). It was founded by Prof. R. Pampuch, now it is headed by his follower Prof. J. Lis. The main scientific directions for them are synthesis of SiC and Si_3N_4 powders, analysis of their structures and properties (chemical, physical and technological ones), sintering and hot-pressing of nonporous compact materials (ceramics), study of ceramics structures and properties, definition of application fields [89,90].

Prof Lis got excellent results in the investigation of the role of Y_2O_3 as a sintering activator. He proved that if the SHS process was carried out in such a way to get a non-equilibrium product (i.e. with a disordered lattice) the influence of the sintering activators is much more remarkable.

The properties of ceramics depend on impurities. But it is difficult to control their content. That is why the ceramics obtained by different companies differ in their qualities. In one of his papers Prof. R. Pampuch presents some kinetic curves for different processes of Si_3N_4 sintering. They proved that the best results were achieved in the case of SHS.

R. Pampuch

Prof. Kashinath C. Patil is a well-known Indian scientist working in the field of combustion chemistry. He develops his works in two directions:

- Surface combustion of redox-compounds obtained from hydrazine and its derivatives;
- Combustion of complex redox-mixtures (Fig. **7.4.1**).

In both cases during the combustion a gas suspension with fine oxide particles (including nano-sized ones) and by-product gases are formed.

Within the first direction Prof. K. Patil and his colleagues synthesized simple oxides, manganites, cobalites, ferrites. Within the second case they obtained refractory oxides, aluminides, zirconates, mullits, cuprates, etc.

The characteristic feature of the working style of Prof. K.C. Patil is combination of synthetic and analytic investigations, aspiration to practical application of synthesized products **[91]**.

In India SHS investigations have become popular. The work is being carried out in many institutions. The specialists of ARCI (Hyderabad) are engaged in synthesizing nitrogen-containing powders based on sodium azide (NaN_3). The Atomic Center in Bombay studies synthesis of intermetallic compounds.

1. Gasless combustion (smoldering)

Initial reagents: $Me(N_2H_3COO)_2$, $Me(N_2H_3COO)_2 \cdot xH_2O$, $Me(N_2H_3COO)_2 \cdot (NH_4)_2$, $N_2H_5Me(N_2H_3COO)_2 \cdot H_2O$	Me=Mg, Ca, Cr, Mn, Fe, Co, Ni, Cu, Zn Combustion: thermal decomposition By-products: NH_3, H_2O, CO_2
Example: $2(N_2H_5)Fe(N_2H_3COO)_3 \cdot H_2O \xrightarrow{air, <1200C} Fe_2O_3 + 6CO_2 + 8N_2 + 16H_2O$	
SHS-products: γ-Fe_2O_3, CeO_2, $MeCo_2O_4$, $MeFe_2O_4$, $MeMnO_4$, ZrO_2, TiO_2, $ZrTiO_4$	

2. Complex redox-reactions in mixtures

Typical redox-mixtures: KNO_3+C+S (gun powder) NH_4ClO_4+Al+CTRB (propellant)	Fuel in redox mixtures: CH_4N_2O, CH_6N_4O, $C_4H_{16}N_6O_2$, $C_4H_4N_2O_2$, $C_2H_6N_4O_2$, $C_3H_8N_4O_2$
Example: $16Fe(NO_3)_3 + 15\,C_3H_8N_4O_2 \rightarrow 8Fe_2O_3 + 45CO_2 + 54N_2 + 60H_2O$	
SHS-products: Refractory oxides, aluminates, aluminum garnets, zerium oxide, zirconium oxide and zirconates, mullites, cordierites, cuprates.	

Fig. (7.4.1): Oxide synthesis with redox-reactions (K. Patil *et al.*).

In the National Laboratory in Jammedpur the SHS study has been organized by an active and very clever woman Suman Mishra (Pattan). She has come to Chernogolovka twice. She studies SHS reactions yielding ZrB_2 and $ZrB_2+Al_2O_3$.

Very interesting results have been achieved by Elazar Gutmanas and Irena Gotman (Haifa, Israel). They study synthesis and properties of various composite materials obtained by volume initiation of SHS processes, i.e. by heating the whole sample in the furnace. This process was called heat explosion by the founder of chemical physics and combustion theory academician N.N. Semenov. These experiments are especially interesting for me because at the beginning of my scientific career I studied heat explosion of explosives and solid propellants and developed heat explosion theory for condensed systems. I studied the macrokinetic aspect of the problem. Our Israeli colleagues develop the material science aspect. Their work is especially interesting because they realize heat explosion in SHS systems under pressure. In this case a compact of powder blends is placed under pressure into preheated pressure cell or between massive preheated rams, rapid heating up to the ignition temperature yields dense products. This approach has been successfully applied to production of various new composite materials [92].

Scientists in Western Europe did not hurry to join us in our work. That is why they have remained behind other colleagues from the other continents. But now they are trying to repair the omission. SHS is investigated in Italy, Spain, France and other European countries.

There are two active groups in Italy; one of them is headed by Prof. G. Cao in the University of Cagliary. He does not have a well-defined interest, so he carries out various investigations within European Projects. I visited the University and delivered some lectures in SHS [93,94].

The leader of the other group is Umberto Anselmi-Tamburini, a follower of Prof. Munir. The main research activity of his group has been focused on experimental investigation of microscopic chemical mechanism of regular SHS processes, modeling of ignition and propagation, influence of mechanical activation and electric fields on the reaction mechanism. I met Dr. Tamburini in 1988 in Prof. Munir's laboratory in Davis. At that time Prof. Munir was finishing his famous review on SHS. He showed me the references and said that 75 % of the list consisted of our papers on SHS. I considered his words as a real praise. But then I had not seen Dr. Tamburini for a long time. I only met him at the IX SHS Symposium in Dijon (France) in 2007 [95,96].

In Spain there is no SHS national organization either. The work in the field started within the international project "Prometheus" which was aimed at trying the capability of SHS under the European terms implying high salaries and specific raw materials and equipment.

I remember the words of one of the Project organizers Mike Werle. He told me: "You see, Alex, the technology in Russia is outdated; it is easy for you to improve it. As to European technologies, they are perfect. If we prove that our technology is efficient under the European terms, they will recognize it".

The project was carried out excellently. The quality of SHS products (a-Si_3N_4 and AlN) was shown to correspond to the World standards, in some cases our products were even purer than their analogs made in furnaces; besides, their cost was much lower.

Soon a large State Company "ENUSA" became interested in SHS technologies. Using the results obtained within the "Prometheus" project, the specialists of the Company developed an excellent automated technology of SHS powder production. Then a small but splendid SHS-powder plant was built. The process of the technology assimilation was slow but at last optimum technological modes were found and the plant could supply high-quality products to the market.

But it had no future. Marketing was started too late, customers were not ready for the new products, besides the Company called "SHS Ceramicas" did not make a compromise with the firms ruling the market. So it was kept out of the market. It had to be closed. It was a useful but very hard lesson for the scientists who thought that the main thing in the business of science-intensive products is a good scientific-and-technical result. It is a pity. According to the opinion of some specialists who visited the plant the level of the automated technology was the highest. There were no such plants in the world. It was unique.

One of the American specialists from Livermore laboratory told me: "We knew that you were building a plant with your Spanish partners. But we were sure that your idea would fail". In Fig. **7.4.2** you can see a photo of the synthesis workshop. In the center there is a robot governing the work. That experience appeared to be a good lesson for us.

We do not have much information about SHS development in Great Britain. Prof. Ivan Parkin, a specialist in oxide chemical synthesis, became interested in SHS and included it to his research methods **[97,98]**. He acquired more experience in SHS due to his collaboration with our researcher Maksim Kuznetsov, a specialist in SHS oxides. Among their results I liked the effect of lattice "densification" best of all.

Fig. (7.4.2): Production line of SHS powders at ENUSA factory in Spain.

There is also a metallurgical company producing some intermetallics and carbides by SHS. But it does not belong to the SHS Society.

Some SHS investigation has been started in Greece. The main specialists there are Dr. C. Agrafiotis and Galina Xanthopoulou (I have already mentioned her) **[99,100]**.

The last who joined us were our French colleagues. They started their work very enthusiastically and created an investigation program under the aegis of CNRS (it is analogous to our Russian Academy of Sciences). Besides they showed a great interest to the collaboration with the Russian scientists. The idea of such collaboration was being discussed during my meeting with Prof. F. Bernard at the VI International SHS Symposium. The main themes of our collaboration are:

- Synthesis of nitride ceramics;

- Mechanochemistry of SHS processes;

- Study of phase formation in SHS processes by using synchrotron radiation and diffractometry.

There are also some other subjects interesting for both sides. The most active person in the development of our scientific contacts is Prof. J.C. Niepce. Our joint work is supported by CNRS representatives and French Embassy in Moscow **[101,104]**.

As to other European countries, the development of SHS there is rather sluggish.

J.-C. Niepce   F. Bernard   J.-C. Gachon

Fig. (7.4.3): Our French colleagues.

7.5. INTERNATIONAL CONTACTS

The beginning of our international activity is connected with the conference on plasma and combustion synthesis. It was organized by Dr. Holt and Prof. Munir within the annual session of the American Ceramic Society.

I was given the honour of delivering an invited lecture at the opening ceremony; it was called "Keynote address". It was clear to me that I should summarize all our achievements. After long thoughts I decided to call my lecture "Self-propagating high-temperature synthesis" twenty years of search and finding". The lecture appeared to be rather substantial though not very long. I had a very good interpreter those days, his name was Sergey Sklyarov, he translated the lecture into English, and as my English was poor I decided to read it.

But we came across some unforeseen consequence. We could not book tickets and did not start our journey in time. So we came to San Francisco very late, it was the second day of the conference. We were extremely surprised to see Dr. Holt and Prof Munir in the airport. They were meeting each flight from New York. It was the beginning of our friendly relations.

But we were unlucky. Our luggage appeared to fly to Miami. But I had my lecture on me. We had a rest in the hotel, and after lunch we were taken to the conference.

The lecture lasted 1 hour 40 minutes (it could be explained by my inexperience, I had never been doing such long presentations since then). I was reading it tensely, pronouncing the words clearly. Everybody understood me. It was very difficult for me to make a lecture in English. I am not fluent in English even now but I can read lectures easily, without embarrassment, confirming the words with interesting illustrations.

The symposium in San Francisco gathered all the outstanding scientists engaged in SHS. I got acquainted with all of them and realized that the world had acquired a good society of scientists keen on SHS.

I was greatly impressed by the first meeting with my foreign colleagues. We discussed what international events should be organized in future and decided to hold SHS symposia every two years. Our decision to hold the first symposium in the USSR was unanimous.

I had a new headache – where to hold the first symposium. I could do it neither in Chernogolovka nor in Moscow. And my old friend Georgy Ksandopulo came to my help. We decided to organize the symposium in Alma-Ata. Prof. N.I. Kidin was joking then: "The Motherland of SHS is Chernogolovka, the Motherland of SHS symposia is Alma-Ata".

The symposium had a success. About 50 foreign scientists attended it. Among them were my new acquaintances from San Francisco. It was 1991. We were discussing where the 2-nd symposium should be held but could not accept the decision. Some months later we got a letter from the well-known American scientist Richard Spriggs who suggested holding the symposium in Honolulu, Hawaii, in 1993. At first we were confused but then certainly agreed. Professor R.Spriggs organized 32 invitations for us and paid our expenses. It was a great wonder. Before the symposium we held the Russian-American Workshop on SHS. It helped Prof. Spriggs to get the money.

I am not going to describe all the symposia – it is worth telling separately. I'd like to mention the places where we held them.

1995 – III SHS Symposium, Wuhan, China, The symposium was held aboard a ship going along the Chang Jiang. A responsible person – Prof. Yuan.

1997 – IV SHS Symposium, Toledo, Spain. Dr. M. Rodriges.

1999 – V SHS Symposium, Moscow, Russia.

2001 (2002) – VI SHS Symposium, Haifa, Israel. Prof. E. Gutmanas.

2003 – VII SHS Symposium, Cracow, Poland. Prof.R. Pampukh.

2005 – VIII SHS Symposium, Cagliari, Italy. Prof. J. Cao.

2007 – IX SHS Symposium, Dijon, France. Prof. J. Bernard.

I am proud that I was Chairman of the Organizing Committees of all these Symposia.

These meetings became very important in our scientific life. Besides we organized bilateral workshops with our partners from different countries. All of them were useful for our cooperation.

When John Kiser came to Alma-Ata, he brought Edward Mikhael Vice-President of the American Publishing House Allerton Press. He had a good experience in cooperation with Russian colleagues as he translated our Soviet journals into English. He liked our discovery and suggested publishing the International Journal "Self-propagating High-temperature Synthesis". We agreed. The journal was issued four times a year since 1992. It appeared to be a good help in our international R&D activity. Soon the number of the papers on SHS was becoming higher and higher, and the journal could not publish all of them. But by the time SHS had become well-known and other journals were ready to publish the articles.

In 2007 "Allerton Press" was purchased by a powerful publishing giant "Springer". It is a new life for our journal with a higher level and stronger requirements. The time will show if we can manage the task. While speaking about our international relations, I'd like to mention the World Academy of Ceramics (WAC). It is a well-known popular organization headed by Dr. Pietro Vincenzini. It sets the pace in ceramic materials science. WAC holds significant forums CIMTEC every four years, organizes various symposia, issues the journal "Ceramics International", rewards its outstanding members with the prestigious International Prize (every four years) for fundamental investigation and innovations in industry. When Dr. Vincenzini learned about SHS, he invited me to give a talk about it. During the next CIMTEC congress we organized a session on SHS (it is a microconference consisting of one or two meetings). Then I was elected an Academician of WAC as well as other representatives of SHS. Among them are J. McColli, Z. Munir, K. Logan, I. Borovinskaya, Y. Miyamoto, R. Pampukh, J. Liss. Now in Chernogolovka there are three Academicians of WAC. The last who was elected was Alexander Rogachev. In 2002–2006 I was the President of the International Advisory Board of WAC.

Dr. Vincenzini and I maintain friendly relations. I rate highly his outstanding organizing relations, and I must admit that I have learned much from him. We organized the Institute of Thematic Associations (ITA) within WAC. The first organization which was included into it was SHS Association (SHS-AS).

Of course our international relations take a lot of time, but no doubt, they are of benefit. The main factor proving the importance of such activity is the existence of friendly community of SHS specialists [42 – 67].

CHAPTER 8

Results of 40-Year Development

Abstract: The Chapter reviews 40-year development of SHS. The main result is the appearance of a new field of knowledge which unites solid combustion and materials science.

What has been done after the discovery of the soilid flame phenomenon and development of self-propagating high-temperature synthesis? Of course, the reader of the book can answer this question himself. Nevertheless, let's "have a walk" along the path from the scientific discovery to industrialization and commercialization of SHS processes. In this chapter we are not going to pay attention to the fact who and where has achieved the result. It will be our common report on the achievements of the SHS community.

8.1. SYSTEMS UNDER STUDY

First of all it was necessary to understand in what systems the phenomenon of solid flame could be realized and what reactions could occur at the SHS mode. At the beginning of our activity we thought that the process was exotic and we wouldn't be able to find many systems for investigation of this phenomenon. Because it required both high thermal effect of the reaction and high melting points of all the substances taking part in the process (Fig. **2.1**). Our apprehension appeared to be unjust. When we realized that it was possible to decrease the combustion temperature by diluting the green mixture with final products without changing the kinetic characteristics of the process, we found more systems for our study. Besides, if we do not restrict ourselves with the requirements of the solid-phase process, it is much easier to find systems for SHS. Of course, we should search them among exothermic reactions because their thermal effect should be rather high so as to avoid significant heat losses into the environment. The terms of self-propagation are defined by combustion theory. But it is impossible to use this prompting constantly because it is necessary to know some values which are not available in reference books due to the complexity of measuring. That is why it is easier to define the terms experimentally. It is a macrokinetic character of self-propagation.

The possibility of the process occurrence in one or another system depends on thermal or kinetic factors. The reaction chemical specific feature is defined by the factors. But the chemical component of the process is rather important for us since it determines the products in each specific case. Fig. **8.1.1** demonstrates the chronology of syntheses in 1967–1990. You can see that each year our work became more and more complicated.

1967	Powder mixtures of metals and nonmetals →	borides, silicides, carbides
	Element-nitrogen systems→	nitrides
1972	Mixtures of metal powers→	intermetallides
	Liquid organic monomers →	polymers
1975	Metal-hydrogen systems	hydtrides
1976	Multicomponent systems→	materials with preset composition
1977	Powder mixtures of metals and S, Se, P→	chalcogenides, phosphides
1979	Mixture of oxides→	complex oxides
1980	Aluminum-reduced systems (SHS)→	refractory compounds + aluminum oxide
1981	Magnesium-reduced systems (SHS) →	refractory compounds + magnesium oxide
1983	Metal-azide systems →	nitrides
1988	Metal – peroxide –oxide-oxygen mixtures →	complex oxides
1990	Mixtures of organic compounds powers →	organic compounds

Fig. (8.1.1): Typical SHS-systems (chronology).

8.2. STUDY OF PROCESSES

A lot of experiments aimed at understanding of SHS nature, regularities and mechanism have been carried out. Methods of experimental diagnostics and theoretical description have been developed. Fig. **8.2.1–8.2.3** demonstrate some diagrams summarizing the results of our long work **[105,106]**.

Using the methods of conventional materials science we studied various combustion products: powders, materials, items and coatings.

Two approaches were used: express-analysis for attestation of SHS products and fine investigation of the product composition, structure and properties.

A new direction was created. It studies dynamics of structure and property formation of final products. As mentioned above it was called structural macrokinetics.

What has been studied well or badly or has not been studied at all? SHS processes in elemental systems have been thoroughly studied. They are the simplest systems. The results obtained during the investigation of Me-B, Me-C, Me-N, Me-Si, Me-H systems (Me – a metal reagent) have become classical.

SHS processes in the systems with one volatile element have been studied less thoroughly; among them are Me-S, Me-Se, Me-P.

SHS processes in condensed systems where both reagents are gasified have not been studied at all. In other words the matter concerns the SHS processes in initially condensed systems in which the reaction occurs in gas phase.

In one of my earliest papers I gave an example of such a reaction: $Mg + 2S \rightarrow MgS_2$

But the problem is still untouched.

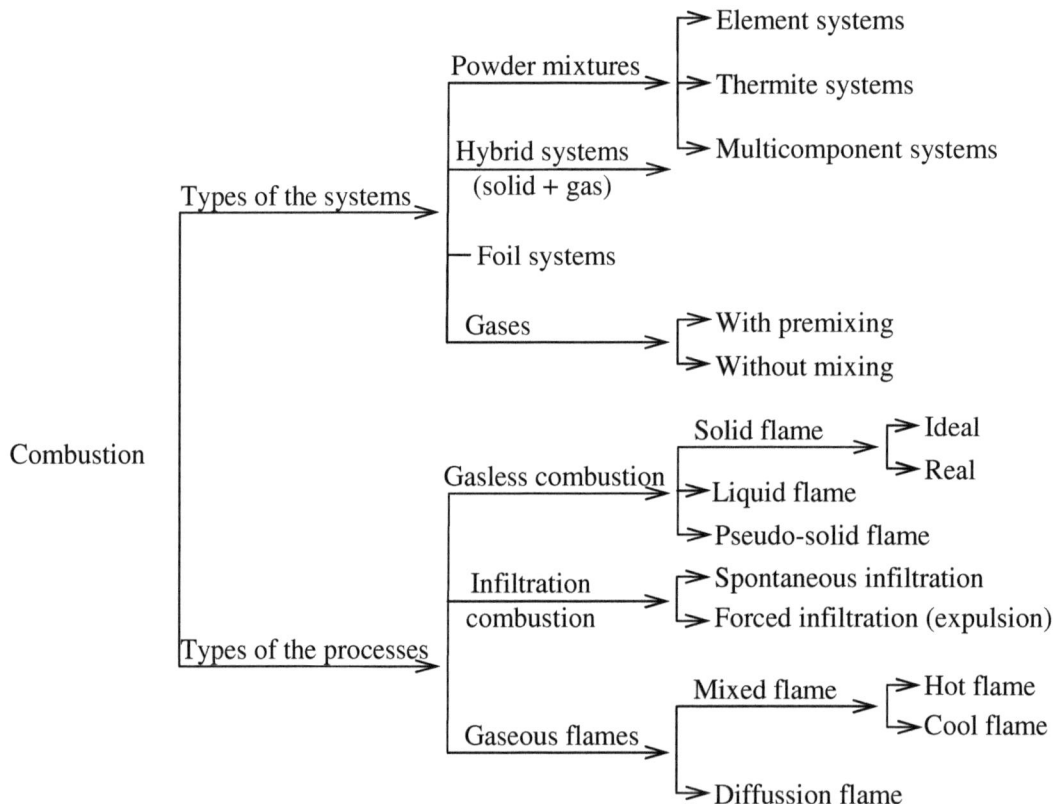

Fig. (8.2.1): Types and mechanisms of combustion.

Phenomenology
(Level 1)

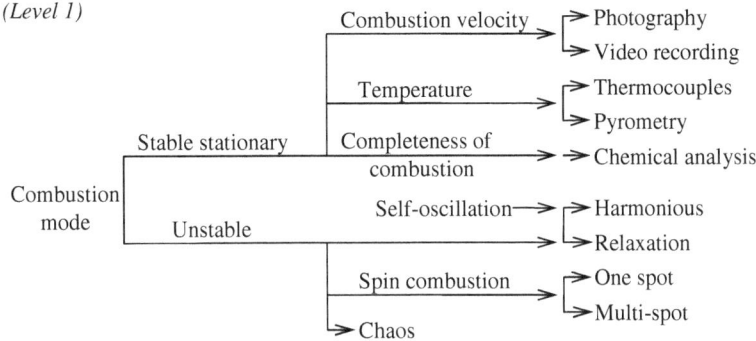

Combustion structure
(Level 2)

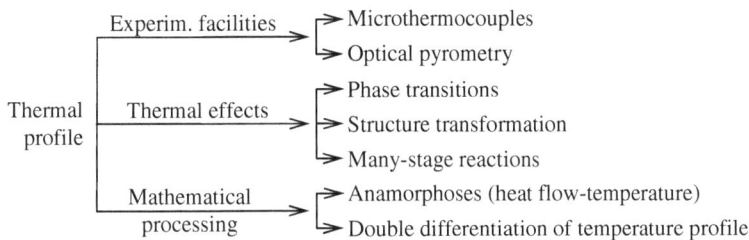

Phase and structure
transformations
(Level 3)

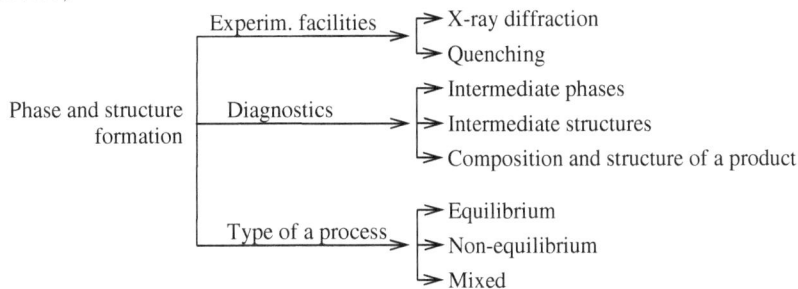

Fig. (8.2.2): Experimental diagnostics of SHS-processes.

In literature on SHS we can find a lot of investigation results of combustion mechanism of more complicated systems. But there are no summarizing studies of the combustion regularities and mechanism. Many specialists, who previously studied combustion of other systems, are engaged in investigation of SHS processes. It provides a high scientific level of the work.

The combustion of SHS systems differs greatly from that of gasifying systems like solid propellants. This difference required new concepts and new approaches.

For instance, phase transitions in the combustion wave are characteristic for SHS processes. But in order to realize their role, it was necessary to study their occurrence in chemically reacting media after fast warming-up. It was found out that there are two regimes of phase transformations: that of Stephan (the phase transformation occurs on the phase boundary, and its velocity depends on the heat flow from the new phase to the previous one) and the volume regime (when the phase transition occurs in a volume which includes the previous phase and a new one, and the velocity depends on that of chemical heat generation in this volume). Therefore, phase transformations influence the wave structure significantly – on the temperature profile of the combustion wave we can see some sharp bends and isothermal plateaus which affect the wave velocity.

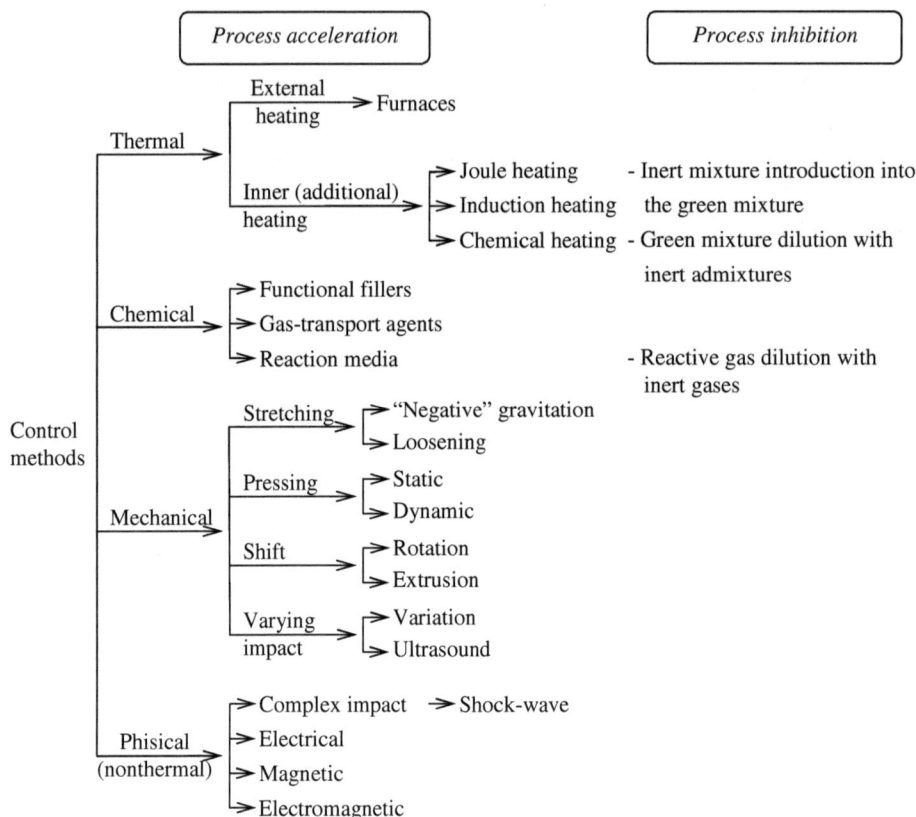

Fig. (8.2.3): Control methods for SHS-processes.

Another example: combustion of SHS systems is also characterized by a high thermal capacity of combustion products. It results in thermal instability of the flat combustion front against longitudinal and transverse disturbances, and consequently, the conventional pattern of the wave propagation loses its stability, and new modes appear (oscillatory combustion, spin waves, chaos).

I think the combustion mechanism and theory in SHS processes is the most successful direction in SHS investigation. A lot has already been done, and it is still an actual problem.

Still at the dawn of SHS I wanted to carry out a gas-phase synthesis. It is the process when partially condensed products are formed in the combustion wave as a result of interaction between gases. The mechanism of the process can be described as follows. An elemental reaction yields atoms of a final solid. While wandering in the flame, the atoms gather in small groups, condense, and form a nano-sized germ, which grows until the entire condensed product of the reaction turns into particles.

I even thought out a name for such a process – chemical-condensation SHS. The fact that it was similar to the well-known "sooting in flames" did not embarrass me, since the future of the process seemed rather rich and promising. After analyzing combustion thermodynamics for some useful reactions, I concluded that in most cases the reaction heat effect was not enough for its self-propagation. But I knew that it was possible to burn gases in electric burners where gas flames are formed due to additional supply of thermal energy. Nevertheless, I could not organize such work.

Nowadays the work in this direction is being carried out, and some useful reactions have been put into practice:

$$2TiCl_4 + 8Na + N_2 \rightarrow 2TiN + 8NaCl$$

$$SiH_4 + O_2 \rightarrow Si + 2H_2O$$

$$SiH_4 + 2O_2 \rightarrow SiO_2 + 2H_2O$$

$$TiCl_4 + 4Na \rightarrow Ti + 4NaCl$$

$$TiCl_4 + 2BCl_3 + 10Na \rightarrow TiB_2 + 10NaCl$$

$$3SiN_4 + 4NaCl_3 \rightarrow Si_3N_4 + 12HCl$$

$$BCl_3 + NH_3 + Q \rightarrow BN + 3HCl$$

Some interesting results are given in Fig. **8.2.4**. The mode of combustion product condensation depends on combination of T_c and P (a pressure in the gas mixture). There can be two threshold regimes: homogeneous condensation yielding nano-particles (aerosol), and heterogeneous condensation yielding nano-films on the surface of the body in flame and on the vessel walls. These data were obtained experimentally under isothermal conditions. The same effects must be observed in flames.

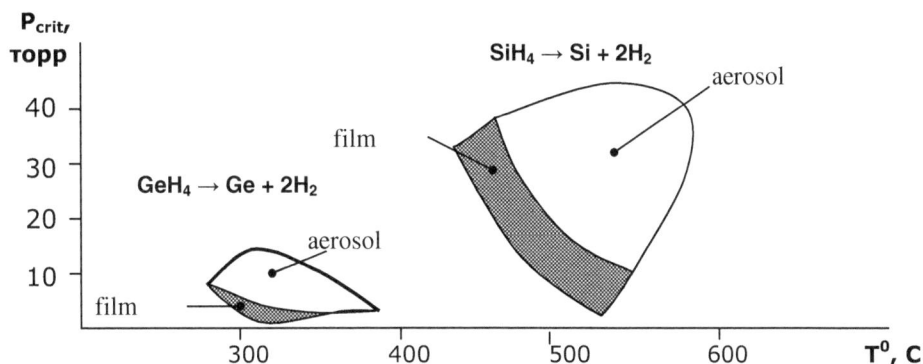

Fig. (8.2.4): Solid product condensation at gas-phase combustion (V.V. Azatyan *et al.*).

I'd like to underline again that it is a very important direction of SHS investigation. It is necessary to extend experimental studies and develop a quantitative theory of the processes.

SHS investigations also deal with structure formation processes, product compositions and properties. The most interesting area here is investigation of dynamics of final SHS product formation. The method of time-resolved XRD analysis (TRXRD) with synchrotron radiation and radiation from an X-ray tube gives great possibilities to study dynamics of a substance's phase composition. But it is not enough. It would be splendid if anybody invented the method for identification of dynamics of other characteristic changes, e.g. lattice parameters, crystallite sizes, porosity, etc. It would be the case when the experimental backgrounds of this new direction appeared in the heart of structural macrokinetics would start sparkling afresh. But they are only my dreams. As to theoretical models of structural macrokinetics, they could have been developed and benefited to the development of the new direction. But I failed to interest the theoreticians in this problem. Perhaps, they will be interested in it after the 40-th jubilee of SHS.

I am not satisfied with investigation of kinetics of heat evolution in SHS processes. It's the main process without which SHS is impossible. When we invented electric thermography, we began studying kinetics of heat evolution for some reactions between metals and gases. I thought that we had ensured the work success. We investigated reactions of Ti, Zr, Nb, Ta with nitrogen, oxygen, and hydrogen. It was done under isothermal terms at a high temperature. We obtained the kinetic equation for the cases and determined the constants. It is well-known that such processes occur under the mode of reaction diffusion, so we can determine the value of the diffusion coefficient.

Then the work was continued with the reactions with complex gases (hydrocarbons, silane). Also we could study the processes of metals' interaction with solid elements because the first stage of the process is pyrolysis with precipitation of the condensed component on the metal surface.

Unfortunately, the work has not been advanced.

8.3. MATERIALS INCLUDING NANO-SIZED ONES

Due to our experience we can conclude that almost any compound produced by a conventional method can be obtained by SHS. There are two ways to do it.

The first way. An exothermal reaction, yielding a preset product, is found. Necessary terms, which provide the required composition and structure, are set up. But there can be one disadvantage – the necessary reagents can be deficient or very expensive and it will make us refuse from the reaction.

The second way (dishonest a bit). We use a conventional reaction but add another highly exothermal one, which yields a useful product. As a result we obtain two useful products by the thermally coupled process.

Thermal coupling can be contact and non-contact. The first method is easier but it makes great demands of the additional reaction (in order to avoid contaminating of the main product with impurities).

Of course, it is clear that it is much easier to choose a proper method among the proposed variety.

Using the SHS method we can solve three chemical-technological tasks.

1. Synthesis of compounds. In this case we make high demands of the composition of the combustion product or the compound obtained after the processing of the product. One of the examples is a synthesis of single-phase compounds of a high purity with a minimum content of reagents, impurities and other phases.
2. Direct synthesis of materials. In this case our aim is to obtain a preset combustion product but not a product after processing. The material structure and properties are of high importance.
3. Direct synthesis of an item. The aim is to obtain items of preset geometrical dimensions, shape and operation characteristics.

The SHS method allows us to solve all these problems. It means that we can obtain a combustion product of a required composition, structure, size and properties. It is the product that satisfies our requirements! Fantastic! But I'd like to note that it is only the beginning of a long way.

Within the first task we have done a lot. We have synthesized a lot of inorganic compounds: simple and complex ones, stoichiometric and non-stoichiometric (including superstoichiometric) phases of a high pressure and temperature, thermally stable and unstable compounds of various types and applications. We can really synthesize them and there are a lot of them. I think now there are about 1000 various compounds. I have written already that at the beginning of our work we studied the possibility of SHS application for synthesizing a compound. The main aim of our work was a synthesis itself but not a product. Then we realized that the SHS opportunities are rather wide and began to synthesize the compounds necessary for our work, i.e. SHS became a preparatory approach in inorganic chemistry.

It is well-known that properties of solid-phase compounds depend on the type and quantity of impurities. Each method brings its own impurities. We know that the compound properties depend on the method of its synthesis. That is why at first the compounds which were obtained by the SHS method were studied as new ones and compared with the available analogs. And there were a lot of misunderstandings.

In spite of the success there are some unsolved problems. One of them is synthesis of non-equilibrium compounds (with a disordered lattice). Such compounds are studied insufficiently.

The synthesis ideology of non-equilibrium compounds is rather obvious. It is convenient to solve the matter by comparing the characteristic times of synthesis – ts, cooling – tc, and thermal relaxation of the lattice – tr. If ts>>tr, we obtain an equilibrium product. If tr>>ts, we have a non-equilibrium product. The cooling time can be high tc>>ts (annealing) and low tc<<ts (quenching). In the case of annealing we usually obtain

equilibrium compounds, and at quenching – non-equilibrium ones.

Everything seems to be lucid. But in practice there are no experimental data, and the properties and application of non-equilibrium products are not quite clear. Some time ago we expressed a simple idea of a dependence of the peak width in the diffraction pattern on the product cooling time and organized some experiments. In Fig. **8.3.1** you can see such a dependence for the combustion product of non-stoichiometric mixture $Ti \pm 0.5C$. The width of Δq peak was measured at its half-height, the cooling time t_c was determined by the combustion thermogram, the dependence $\Delta q(t_c)$ was a drooping curve with saturation. Since the peak width is an indirect characteristic of the crystal lattice order, it is obvious that the points on the drooping branch corresponds to the formation of its non-equilibrium state and on the saturation section – to that of the equilibrium state. This result led us to the important conclusion that during the product cooling its auto-annealing occurred (it is usually carried out at a constant temperature but not at its decreasing value as in SHS processes). It is possible to obtain either equilibrium or non-equilibrium combustion products by regulating the cooling time.

I am sure that such work should be continued but with specialists in substance structure in order to characterize the non-equilibrium state of the lattice more thoroughly.

Another example is connected with synthesis of organic compounds. Their formation temperature is much lower than that of inorganic compounds so you can ask me a question if we should carry out the organic synthesis under the SHS mode, since the advantage is insignificant. Besides, the temperature which is required for organic SHS can be easily achieved in furnaces.

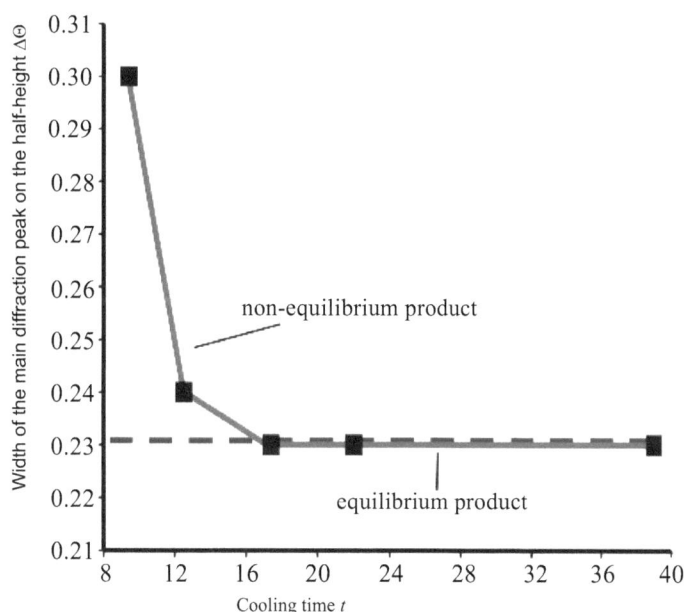

Fig. (8.3.1): Transition from non-equilibrium state to equilibrium one (A.G. Merzhanov, V.M. Shkiro, D.Yu. Kovalev) for $Ti + 0.5C \rightarrow TiC_{0.5}$.

From my point of view there are three reasons to study organic SHS **[107,108]**.

1. The temperatures and velocities are not very high that is why this direction is of interest from the viewpoint of experimental diagnostics of SHS processes. Besides, some new investigation methods developed within this direction can be used for organic systems.

2. Sometimes it is convenient to synthesize organic compounds from powders (by the "dry" method, not the "wet" one in solutions).

3. The technological efficiency of SHS in this case is not high but the product output is huge.

Fig. (8.3.2): Organic SHS: regularities of wave propagation.

Fig. **8.3.2** demonstrates the combustion regularities of piperazine-malonate synthesis and in Fig. **8.3.3** you can see some examples of organic reactions. This direction is rather promising. But since 1990 only one researcher has been studying this problem.

Reactions	Parameters	
	U, mm/sec	T_m, °C
Oxidation $C_6H_6O_2 + KBrO_3 \rightarrow C_6H_4O_2 \cdot C_6H_6O_2 + KBr + H_2O$ hydroquinone quinhydrone $C_6H_6O_2 + KBrO_3 \rightarrow C_6H_4O_2 + KBr + H_2O$ quinine	0.1	160

Protonation
    $C_4N_2H_{10}$ + $C_3H_4O_4$ → $C_4N_2H_{10} \cdot C_3H_4O_4$
    piperazine   malonic acid   piperazine malonate
	1.4	130

Halogenation
  $C_3H_4O_4$ + $C_4H_4O_2NBr$ → $C_3H_3O_4Br$ + $C_4H_4O_2NH$ → $BrC_2H_3O_2$ +
                      CO_2
Malonic   N-brom-   Brommalonic acid   Bromacetic acid
acid   succinimide
   $C_3H_4O_4 + C_4H_4O_2NBr \rightarrow C_3H_2O_4Br_2 + C_4H_4O_2NH \rightarrow Br_2C_2H_2O_2 + CO_2$
         Dibrommalonic acid   Dibrommalonic acid
	1.0	90

Oxidation-halogenation
  $C_9H_7NO + C_6H_5SO_2NNaCl \cdot 3H_2O \rightarrow C_9H_7NO_2 \cdot H_2O + C_6H_5SO_2NH_2 + NaCl + H_2O$
8-   chloramine B   8-oxyquinoline
oxyquinoline   N-oxide hydrate
  $C_9H_7NO + C_6H_5SO_2NNaCl \cdot 3H_2O \rightarrow C_9H_6NOCl + C_6H_5SO_2NH_2 + NaOH + 2 H_2O$
 	
                    Cl-substituted
                    8-oxyquinoline
	0.5	110

Oxidative imitation
  $(C_6H_5)_3P + C_6H_5SO_2NNaCl \cdot 3\ H_2O \rightarrow (C_6H_5)_3P{=}NSO_2C_6H_5 + NaCl + 3H_2O$
                    triphenyl-N-(phenylsulfoyl)-
                    phosphinimine
  $(C_6H_5)_3P + C_6H_5SO_2NNaCl \cdot 3\ H_2O \rightarrow (C_6H_5)_3PO + C_6H_5SO_2NH_2 + NaCl + 2H_2O$
 	
                    triphenyl- phosphineoxide
	4.9	240

Acylation
  $(C_5H_5)_2Fe$ + $C_8H_4O_3$ $\xrightarrow{AlCl_3}$ $C_5H_5FeC_5H_4COC_6H_4COOH$
  ferrocene   Phtalic   o-
            anhydride   carboxybenzoylferrocene
	0.5	185

Fig. (8.3.3): Organic SHS: reactions and products.

Now about materials.

The most popular type is powders. Some specialists do not consider them materials and always say "powders and materials", but others think that powders are also materials and call them powder materials. We do not care and say "SHS powders".

There are different types of SHS powders: agglomerated, single-crystal, and composite ones.

The powders are usually obtained by processing combustion products (grinding and chemical and thermal treatment). Each SHS product can be processed into powder. The main characteristic of SHS powders is their morphology which is various (Fig. **8.3.4**) and allows us to understand its influences on the powder properties. The achievements of SHS in this field are well known. Many organizations have their own pilot workshops where they produce small lots of SHS powders. The typical particle size is 1-5 mm (crystallite size). The combustion product is usually ground in such a way to disintegrate agglomerates and obtain nonporous single crystal powders.

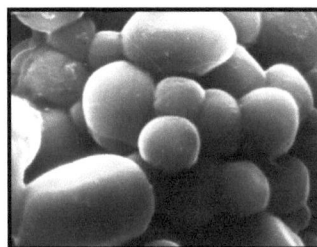

**spherical particles
(low-temperature AlN)**

**columnar grains
(high-temperature AlN)**

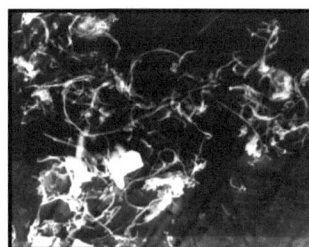

whiskers

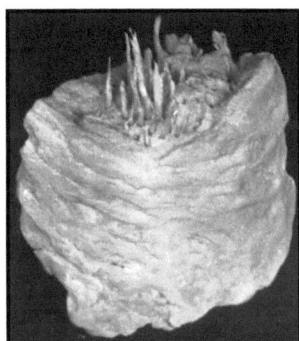

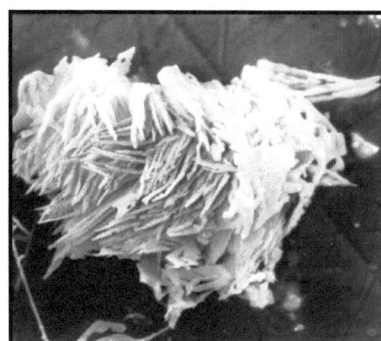

Fig. (8.3.4): SHS powders (I.P. Borovinskaya et al).

We have never striven to obtain submicron powders (0.1–1.0 mm) as it is difficult to select proper sintering terms for them due to their significant shrinkage. Furthermore, they are not used for gas-thermal coatings. But suddenly the problem of nano-sized materials arose. They include powders with nano-sized particles, compact materials and coatings with nano-sized crystallites, porous items with nano-sized pores. The appeal to develop nano-sized materials came from the governments of all the countries with highly developed materials science. The officials of high rank but incompetent in science headed various programs, councils, committees and so on. They promise a better life due to nano-materials and nano-technologies. And what about scientists? Each of us tries to steak this prefix to any word; some even contrive to do it in their annual reports. Everybody is hurrying. Where? To snatch a piece of a nano-cake…

As to SHS scientists who did not want to deal with fine particles and grains, they appeared to synthesize nano-materials, especially nano-powders, without any efforts but did not make a show of it because they did not know how to use them.

I am sorry for such frivolous comments on this very important problem.

Speaking seriously, SHS has been thought to be inapplicable to synthesis of finely dispersed powders because of high temperatures which cause recrystallization enlarging grains. But we can struggle against it.

There are a lot of methods preventing recrystallization. Fig. **8.3.5** shows us the main methods which allow obtaining SHS nano-powders [109,112].

Conventional approach	SHS stimulated by carbon combustion	Gas-phase SHS	SHS with high-temperature dilution	Solution combustion	SHS in multi-layer nano-sized films
Recrystallization suppression yielding nano-sized crystallites		Formation of gas dredge with nano-sized particles	Combustion of SHS mixtures diluted with high-temperature powder agent	Mixture of nano-sized powders	
I. Borovinskaya V. Vershinnikov (Russia)	D. Luss (Italy) K. Marti-rosyan (USA)	R. Akselbaum (USA) V. Azatyan (Russia)	I. Borovin-skaya (Russia) S. Mishra (India)	K. Patil (India) A. Mukasyan (USA)	A.Rogachev, A. Grigoryan (Russia)
The most assimilated direction – oxide production					

Fig. (8.3.5): SHS among nano-technologies: various approaches.

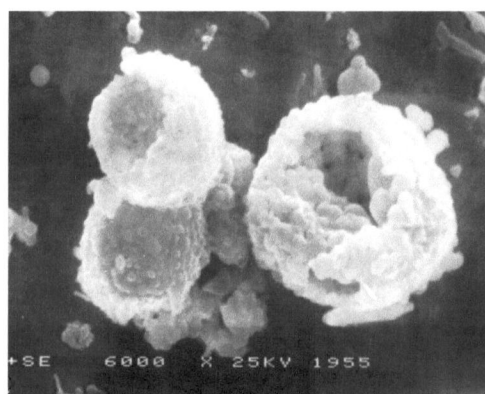

Single-crystalline powder,
Grain size ~ 0.1μm
(complete disintegration)

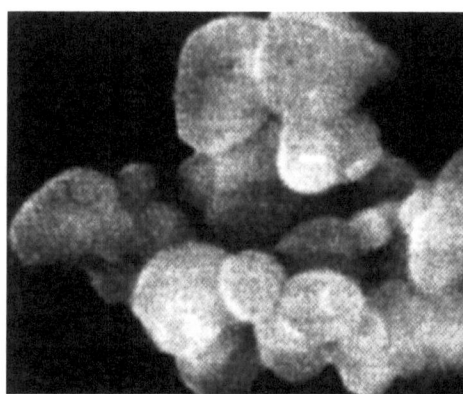

Hollow agglomerated particles
(partial milling)

Fig. (8.3.6): First nano-particles obtained by SHS – AlN powder (I.P. Borovinskaya, N.S. Makhonin, L.P. Savenkova, 1991).

I'd like to underline that nano-sized particles can be obtained by two methods:

- By milling large crystallites (e.g. in attritors),
- By carrying out the process in such a way to obtain nano-sized crystallites.

I think that only the second method is of interest for this problem. The crystallites' properties differ greatly from those of conventional particles.

But it is important for us to go beyond the synthesis. The nearest tasks of the SHS specialists are to gain experience in analyzing nano-sized powders and to determine the efficient ways of their application.

If we want to occupy our own niche in this area we should study the preliminary structure formation of crystallites, find new approaches regulating their sizes, and analyze the influence of nano-effects on material properties. It's a very important fundamental task.

Now about compact materials. There are two ways to obtain them:

1. From SHS powders using the methods of powder metallurgy (sintering, hot pressing).

2. Directly during SHS without the powder synthesis stage (direct synthesis).

In the first case SHS supplies raw materials. The main work is carried out by powder metallurgy. But SHS powders differ greatly from their analogs produced in furnaces. So there were some difficulties on the path to their assimilation. I'd like to give you an example connected with SHS titanium carbide. Our elaboration of abrasive pastes based on SHS-TiC was a success and we decided to use it for making tungsten-free hard alloys. We suggested the idea to the USSR Ministry of Non-ferrous Metallurgy and sent them a pilot lot. Soon they answered that SHS-TiC was not applicable for hard alloy production. We wanted to know – why? And soon I clarified that our powder had been sent to the head organization engaging in hard alloy production. They ground it and then sintered according to their conventional method. They obtained a solid highly porous sample.

I realized that before the conclusion they should have done some investigation, changed the grinding and sintering methods and found optimum terms of the technological operations. They had not done it. It was their mistake. But I did not want to argue.

Some time later we returned to this problem. A well-known specialist in materials science from Leningrad Polytechnic Institute Sergey Ordanyan charged his post-graduate student (it's a pity I don't remember her name) with investigation of this case. She proved that the structure of our titanium carbide was more rigid that that of the conventional one; therefore, the ties between the crystallites were stronger. So the specialists from that organization had sent undermilled product to sintering and got such a bad result. That post-graduate student also studied the dependence of the sintering mode on milling and showed that titanium carbide had to be milled to complete disintegration of its crystallites. She found the optimum sintering mode, made samples of the hard alloys, and measured their hardness and shearing strength. She proved that SHS-TiC is one of the best materials which can be used for making hard alloys because its strength was 10-20 kg/mm^2 higher than that of the conventional titanium carbide. I wanted to explain the mistake to our specialists from that leading organization and made a diagram for them (a very simple one).

The essence of this plot is the following: a new material can have worse properties in comparison with an old one if it is sintered by the same method. In order to get high operation characteristics one must study the dependence of the characteristics on the process parameters and find optimum terms. And only then the characteristics of the products obtained under these optimum conditions should be compared. Of course we were not sure that our SHS product would be better. But then we could see that in most cases the SHS product is better than its analog obtained in furnaces. We explained it as follows: the SHS processes is a multiparameter one, it means that we can regulate the process more easily. But in spite of our arguments and achievements the Ministry insisted on their preliminary conclusion.

Later on we made hard alloys based on SHS titanium carbide with molybdenum carbide. Among tungsten-free hard alloys they appeared to be the best from the viewpoint of their characteristics (Fig. **8.3.7**). But the time was lost. We started developing new problems and did not return to that one.

$Ti_xMo_{1-x}C_x$	Composition, wt. %			
	Mo	C $_{bound}$	C_{free}	O
0.93	11.4	17.9	0.1	0.2
0.95	8.1	18.0	0.05	0.2
0.97	4.8	18.5	0.2	0.3

$Ti_xMo_{1-x}C_x$	Composition TiC, wt. %		Hardness, MPa	Bending strength, MPa	Operation stability coefficient at cutting, K	Tool stability at diamond synthesis (number of cycles)
	Carbide	Binding element (Ni)				
0.93	84	16	91.5	1270	1.4	2300
	88	12	91.7	1080	1.6	2100
	92	8	91.9	1010	1.8	1900
0.95	84	16	92.3	1520	2.1	3560
	88	12	92.6	1480	3.0	3760
	92	8	92.7	1420	3.1	4650
0.97	84	16	92.4	1800	2.5	6980
	88	12	92.2	1570	2.7	4160
	92	8	92.3	1400	3	3240

Fig. (8.3.7): SHS hard alloys based on titanium carbide alloyed with molybdenum carbide.

Nowadays a lot of materials with various characteristics for different industries are obtained by the SHS method [113,114]. You can see some of them in the Table below (Fig. **8.3.8**). I'd like to dwell on so called "black ceramics" (Si3N4+SiC+TiN) characterized by a very low friction coefficient. It has some peculiarities:

1. It is synthesized by burning the green mixture in high-pressure nitrogen.

2. During the combustion the shape and sizes are practically the same.

3. It is possible to obtain samples of a low residual porosity (~1%).

4. The sample surface is smooth, i.e. machining is not required.

5. Bending strength does not depend on temperature.

6. The friction coefficient of black ceramics is very low.

The ceramics can be used for tribotechnical purposes. But it still needs some theoretical study. The items made of black ceramics are shown in Fig. **8.3.9**.

Now a few words about synthesis of items [115,117]. It is one of the most complicated tasks. Three technological types (TT) allow obtaining items of a preset shape and size:

-   SHS sintering (TT-2): an initial mixture is shaped as an item to be sintered. Combustion is organized in such a way to preserve the sample shape and size. The combustion product needs no machining and its porosity is 5–15 % (black ceramics with ~1% residual porosity is an exception). This method is used for manufacturing ceramic insulators for furnaces of oriented crystallization.

-   Forced SHS compaction (TT-3): in order to obtain an item of a preset shape and size the combustion product should be pressurized in a mould of the same internal shape and size. It is possible to be done but there are still a lot of problems, e.g. temperature heterogeneousness during cooling. This method is still under study. If it can't be used, we can apply another method – a billet is made and then a goal item is cut out of it.

-   The technology of high-temperature melts (TT-4). High-temperature moulding is not an easy task. It is difficult to use the experience of metallurgy because of a significant difference in temperatures. But if the obtained melt remains inside the chamber used for the process, the task becomes easier. An example of such an approach is the above mentioned centrifugal technology of pipe production.

7. So this revolutionary process is only at the beginning of its development.

Composition	Porosity	Destination	Application
BN	30–50	Rods, crucibles, bushings, bricks, plates, fixing items	Electroinsulating corrosion-resistant bushings; crucibles for ferrous metal and amorphous alloys; lining of high-temperature heaters
B-BN $B-BN + Me_mO_n$	~ 15	Lining bricks	Biological protection
$BN + SiO_2$	~ 10	Lining bricks, bushings	Lining of MGD-generator channel; erosion-resistant burner bushings for air-plasma cutting of metals and alloys
$BN + TiB_2$	~ 10	Lining bricks and plates, metal lines, stop valves, steel-pouring nozzles	"Pouring" amorphous tapes; biological protection; spray powders for tool and structural stainless steels
AlN	20–30	Cylinders, plates	Sintering fixture for heat-conducting boards of integrated circuits
$AlN + TiB_2$	<5	Bricks, plates for evaporation elements	Lining of electrolysis baths; deposition of aluminum coating
SiAlON	20	Items of a complicated shape: turbines, items with honeycomb structure, lining bricks and plates	Parts of internal combustion engine for turbo blowing; catalyst carriers for incineration exhaust gases
SiAlON + SiC + BN	<10	Turbines for internal combustion engines, lining bricks, steel-pouring nozzles	Parts of internal combustion engine, electrolysis bath protection (in Al production); metallurgical fixtures
$SiAlON + TiB_2$	~ 15	Lining bricks and plates	Parts for metallurgical industry, black ceramics
$Si_3N_4 + SiC + TiN$	<5	Plates, rods, bushings, spherical items	Friction pairs, wear-resistant parts for internal combustion engines, ball bearings
$Si_3N_4 + SiC + TiN + C$	10–15	Bushings, plates	Parts operating at high temperatures
SiAlON-BN	20–30	Lining bricks and plates	Items for metallurgical production

Fig. (8.3.8): SHS ceramic materials.

8.4. SHS: TECHNOLOGY, EQUIPMENT, PRODUCTION

In chemistry they usually start with a laboratory experiment and then proceed to industrial production. I do not remember exactly but I think that the technological stages were defined in chemistry:

1. A laboratory technology.

2. A pilot-scale facility.

3. Semi-industrial production.

4. Industrial-scale production (usually automated production lines).

What does it mean?

1. Obtaining a principal result and a small portion of a product in a laboratory and then searching for partners for further common R&D activity.

2. Developing the technology at the author's laboratory and producing small lots of the product to define its application fields, organize preliminary marketing and evaluate the product efficiency.

3. Production for need of a given (permanent) customer. It is usually located at a remote site (often a large factory) as its individual division. Normally, such a production is completely or partially automated. The main objectives are Industrial-level checking of the process with the elements of business.

4. A large production output and realization of the product on the market. As a rule, this is a set of automated production lines that forms an individual plant (company).

From the viewpoint of R&D activity, SHS has one very important peculiarity: the work according to items 1–3 is carried out using the same equipment and technological scheme and differing in its aim and output.

Density, g/cm^3	3.40
Bending strength, MPa (up to 1500°C)	650
Modulus of elasticity, GPa	300
Hardness, HRA	93.5
Hardness, H_v	20.0
Friction coefficient - dry friction at 273 –1173K - hot friction	0.2-0.3 0.02-0.03

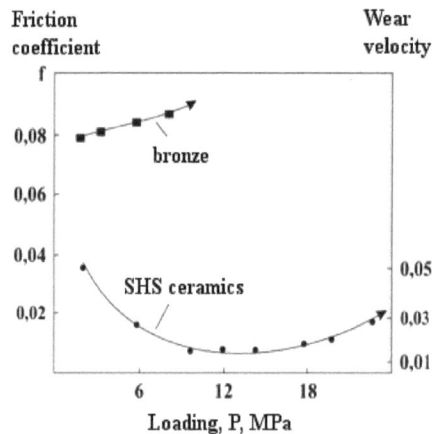

Some machine parts made of black ceramics:

1	–	pusher;
2	–	pusher bushings;
3	–	bushings of an intermediate gear;
4, 8	–	friction pairs;
5	–	ball-bearings;
6	–	tips of busher bar;
7	–	wear-resistant plates

Composition: $Si_3N_4 – SiC – TiN$ (47-50%, 26-27%, 15-17%)

Fig. (8.3.9): Characteristics of "black" ceramics (I.P. Borovinskaya, M.Yu. Blinov, A.S. Mukasyan).

For example, nowadays, if we have a synthesis reactor of 20 l capacity, it will take us about 3 or 6 weeks to develop a technological process without carrying the corresponding laboratory experiment. To be more exact, we do not need a laboratory technology in many cases. Another example: we have a press of 2 thousand tons for making large-scaled hard-alloyed items by SHS compaction. Usually it is necessary to carry out several syntheses to obtain a high-performance product. The question is – What kind of technology is this: a laboratory, pilot-scale or semi-industrial one? Probably, it is a lame example because the items obtained with such a press are made by request.

I'd like to underline that the semi-industrial facilities which were organized by "Termosyntez" (Fig. **5.3.1**) had tested the SHS technology and proved its reliability before the USSR disintegration. As to industrial technological lines which are completely or partially automated, it is beyond my competence. But I can conclude that the technology (either a pilot-scale or semi-industrial one) is characterized by high efficiency and can compete with the conventional furnace technologies.

Let's list the results of our activity in this field.

- Technological Regulations and other technological documents have been developed almost for all powder and direct synthesis technologies.

- Pilot-scale facilities have been organized for producing various SHS products.

- Various types of SHS equipment have been designed and fabricated: universal reactors for powder synthesis, special moulds for forced SHS compaction and SHS extrusion, centrifugal SHS chambers, constant pressure vessels (it is original equipment).

- Semi-industrial facilities have been organized in some plants.

- Technical and cost efficiency of the SHS technology and products have been recognized.

- Independent SHS plants have been founded (there are only few but they are working). Some of them are equipped with automated technological lines.

- In some Russian universities and institutes the lectures in SHS theory and practice are delivered.

Therefore, SHS can be regarded as a challenge to other power-consuming processes for synthesis of inorganic materials.

The characteristic features of the technology are:

- utilization of cheaper chemical energy instead of electric power;

- simple equipment due to the absence of external heat sources;

- high process velocities connected with significant self-heating in the combustion wave;

- layer-by-layer heat evolution and the possibility of working with a huge quantity of a substance;

- possible organization of continuous SHS production.

Some conventional processes and their SHS analogs are given in the Table below (Fig. **8.4.1**).

Conventional processes of inorganic material production	Alternative processes
Furnace and plasma-chemical synthesis	SHS (TT-1)
Sintering and hot-pressing	SHS sintering (TT-3)
Plastic working	Forced SHS compaction (TT-2)
Gas-thermal spraying	Spraying using thermally reactive powders
CVD-processes, diffusion saturation	Gas-transport SHS technology (TT-6)
Metal-thermal smelting	SHS metallurgy (TT-4)
Electric-arc surfacing	SHS surfacing (TT-4)
Electric welding	SHS welding (TT-5)

Fig. (8.4.1): SHS as an alternative technology.

But our R&D activity in the SHS field has clearly demonstrated that it is too difficult to put a new technology into practice. You are always confronted with groundless objections. It seems that nobody thinks about the country's interests – private interests and ambitions are put in the forefront. It was necessary for us to prove our rightfulness. But it is natural: the "old" already exists and works and we must try to find a place for the "new". We did our best. We emphasized the SHS technical and cost efficiency.

It is not difficult to analyze technical efficiency. One must know the most important characteristics of the old product (they are available) and the new one (they should be measured). On comparing these characteristics, one can realize which product is better.

The data on cost efficiency are not usually available. It is a secret of each enterprise. If the director is not interested in the introduction of a new technology, it would be impossible to evaluate the economic efficiency. So, one of the most significant tasks of the author is to interest the administration of the enterprise in this elaboration. Sometimes, prescient scientists invite some executives to be co-authors of their innovation.

Though the evaluation of cost efficiency a priori can have serious errors, we tried to get the required data and analyze them. Figs. **8.4.2**–**8.4.4** demonstrate our results in this direction, and frankly speaking, some of them are rather impressive.

Thus, we developed a plan of possible actions from an engineering idea to industrial assimilation (Fig. **8.4.5**) **[118]**.

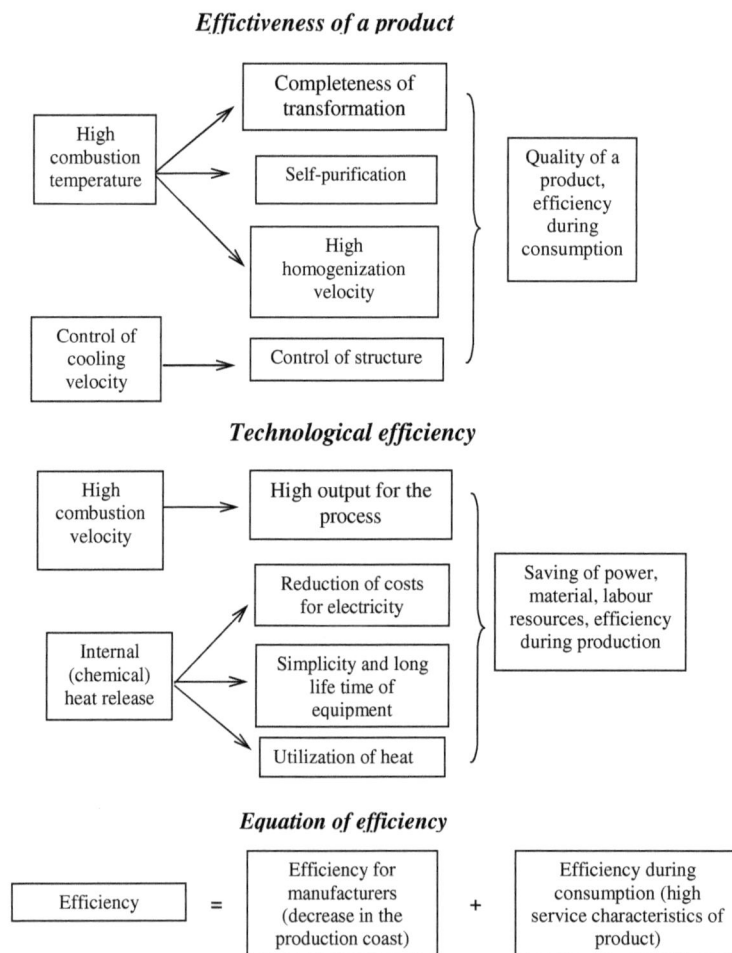

Fig. (8.4.2): Physical basis of efficiency.

| SHS technology | Ration of SHS product cost to the analogs' prices in the market | Cost share, (%) | | | Man hour for 1 kg product | Power consumption for 1 kg product |
		Raw material	power	labor		
Cast powders (chromium carbide, etc)	0.25–0.35	40–45	0.2–0.5	13–18	0.15–0.25	0.6–0.7
Elemental powders of non-metal nitrides (aluminum and silicon nitrides)	0.13–0.25	25–30	0.8–1.0	11–15	0.06–0.09	0.8–1.2
Composite powder ($TiC-Cr_3C_2$) + Ni	0.27–0.32	65	0.4	9.7	0.13	0.65
Magnesium-redused powders (titanium diboride, boron nitride)	0.4–0.45	60–80	1.6–1.8	1.5–1.7	0.88–0.93	5.6–6.2

AlN production	SHS	Furnace synth.	Furnace-chemical synthesis
Aluminum consumption, kg/kg	0.7	0.9	1.5
Nitrogen, m^3/kg	0.9	1.65	12.3
Power consumption , kW h/kg	0.5	31	150
Labor consumption, rel.un.	1	1.4	3.5
Number of technological stages	8	18	5
Synthesis installation output, Kg/h	4	1.0	0.75
Synthesis time, h	0.6	2.5	0.5
Powder cost, rel.un.	1	2	4

Fig. (8.4.3): Some technical-economical data of SHS technology and efficiency of AlN production by three different methods.

| SHS product | Commercial analog | Improvement of technical parameters | | Other SHS advantages |
		parameter	Improvement	
Abrasive material based on pink corundum	Synthetic corundum	Increase in polishing ability	1.4–2.3 times	Increase in microhardness (1.2 times)
Titanium-chromium-nickel composite for coatings	Titanium carbide produced in furnaces	Increase in wear resistance	2.8 times	Used as an abrasive material
Silicon nitride powder as a stuff of dielectric sealer	Silicon nitride produced in furnaces	Improvement in dielectric parameters	1.2 –2.0 times	High thermal conductivity of a support
Magneto-soft ferrite powders (Ni-Zn, Mn-Zn)	Ferrites produced in furnaces	Decrease in scattering of electro-physical parameters	2 times	Decrease in production cost (1.5–3 times)
Granular material based on Ti-Cr-B for spraying	Melted tungsten carbide	Power economy at production	250 times	Decrease in labor consumption (1.7 times)
Molybdenum disilicide powder for electric heaters	Heating elements made of molybdenum	Increase in thermal stability	14 times	Decrease in production cost (5-10 times) Increase in ecological safety

	disilicide produced in furnaces			
Nitrided ferro-vanadium for alloyed steel	Ferro-vanadium nitrided in furnaces	Increase in nitrogen content	2 – 3 times	Increase in nitrogen assimilation by steel
Zirconium aluminide as getter in incandescent lamp	Zirconium powder	Increase in operation life	1.2 times	Decrease in production cost (2.9 times)
Molybdenum and tungsten disulfides as materials for dry friction	Natural molybdenite	Increase in operation life of friction pairs with disulfides	10 times	Quality improvement
Fumigant "Termofos"	Fumigant "Fostoksin" (Germany)	Increase in phosphorous hydrogen release	1.5 times	Ecological safety at synthesis
STIM-5 unreground cutting plates	Cutting plates made of T15K6 hard alloy	Increase in operation life	3 times	High durability of cutting elements
Resistive targets for film magnetron evaporation	PS-4800 resistive alloys	Fatigue characteristics after 1000 hours	10 times	No machining after synthesis
Large billets for rolls	Billets made of M-13 steel (USA)	Increase in durability	2.5 times	Improvement of rolling
Items with corrosion-resistant coatings	Uncoated items	Increase in operation life	3 – 5 times	Coating thickness 0.15–0.2 м. Microhardness 8000–24000 MPa

Fig. (8.4.4): SHS products.

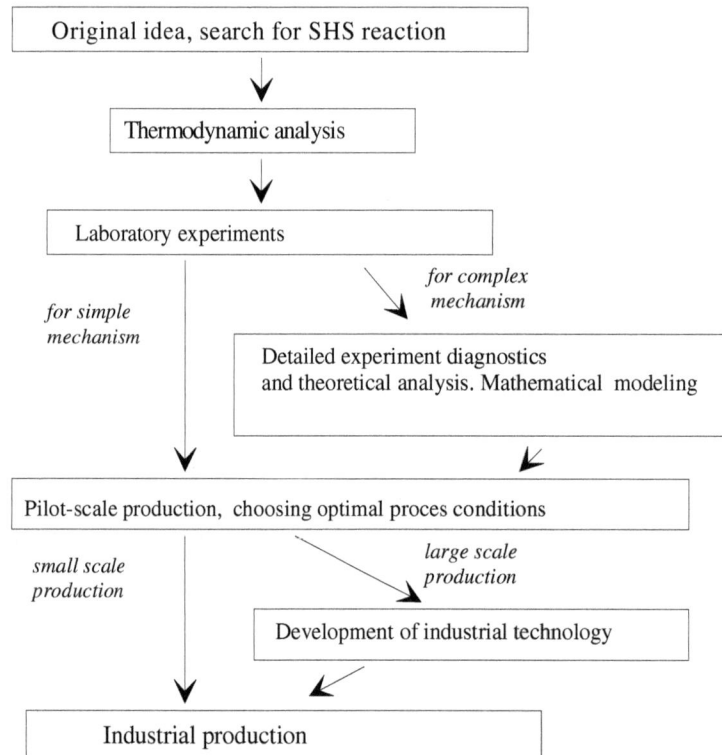

Fig. (8.4.5): From initial ideas to industrial production.

But I'd like to underline that assimilation of some elaborations is very slow and unexpected.

Some years ago I published the article "SHS on the pathway to industrialization" where I set forth the ideology and state of art of our investigations and also gave examples of the most impressive production facilities. But now I'd like to tell you about two new Russian facilities. One of them was founded in the town of Magnitogorsk for manufacturing composite alloys to be used in melting steel and cast iron. The main product was nitrided ferro-alloy. This work had originated in Chernogolovka, in I.P. Borovinskaya's thesis dedicated to combustion in nitrogen and to nitride synthesis. Then this work was continued in Tomsk and headed by Yu.M. Maksimov and his follower Mansur Ziatdinov. Then Mansur started working in this direction in Magnitogorsk. His colleagues have created a joint company and developed a large-scale industrial technology of nitrided ferroalloys and base metals (Fig. **8.4.6**). You can see some of its parameters:

Workshop area – 1200 m^2,

New reactor capacity – 150 l,

One-time loading – 0.5 tons,

Production output – 2 400 tons

Combustion heat utilization – the workshop heating.

I have never seen such huge reactors.

Frankly speaking, it is not a complicated task to choose the volume of a reactor. But it depends on the properties of the material to be synthesized.

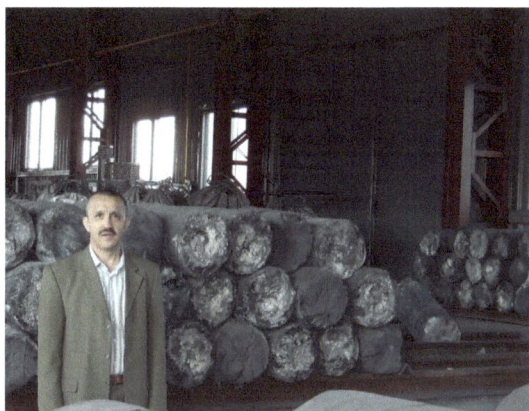

Fig. (8.4.6): M. Ziatdinov, the organizer of industrial production of nitrided ferro-alloys in Magnitogorsk.

The activity of this workshop is an excellent reply to skeptics who consider that such large facilities based on SHS cannot be organized. It is also praise to Yury Maksimov for the development of this SHS technology, to our designer Victor Ratnikov, technologist Valentina Prokudina and Head of the Laboratory Inna Borovinskaya.

I am sure that this is a good example for those who want their results to unite people in the cause of science and new technologies.

The second facility is being built in Chernogolovka. It is a plant of powder metallurgy of tungsten carbide. A semi-product is obtained by SHS according to the following scheme:

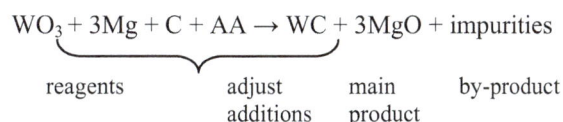

$$WO_3 + 3Mg + C + AA \rightarrow WC + 3MgO + \text{impurities}$$

| reagents | adjust additions | main product | by-product |

We managed to define the terms when the other phase W2C does not form. The combustion product is subjected to acid enrichment when all the impurities are removed. It seems to be rather simple. But the additives which we have to introduce into the green mixture are removed badly.

Parameters	VK6-OM		VK8-OM
	WC alloy	Standard	WC alloy (SHS-M)
Specific weight, g/cm^3	14.9	14.7	14.6
Hardness, HRa	91.0	90.5	90.0
Bending strength, kgf/mm^2	170.0	130.0	164.5
Stability coefficient	1.4	1.0	
Porosity	0.04	0.2	0.2
	"Pobedit" Plant		"Virial", St. Petersburg

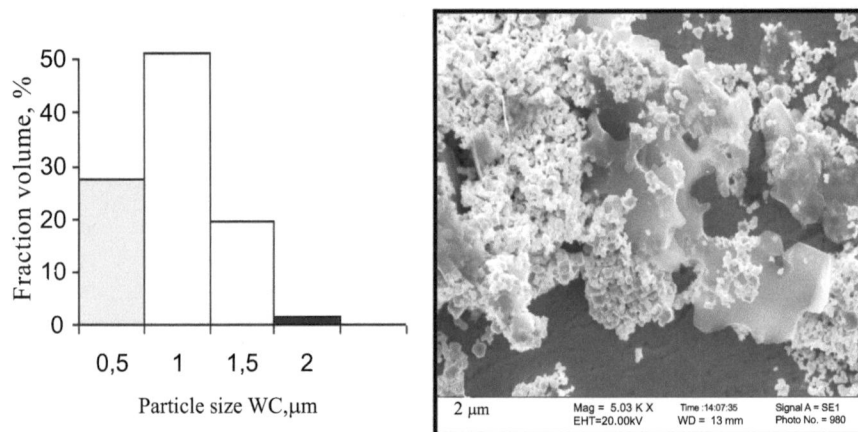

Fig. (8.4.7): Synthesis of submicron WC powders with a reduction stage.

As a result of our work a submicron powder WC was obtained. The powder and hard alloys made of it have very good characteristics (Fig. **8.4.7**). The task was solved by a friendly team which united V.I. Vershinnikov (the main executor of the work), T.I. Ignatieva (chemical processing), A.V. Goziyan (a supervisor), I.P. Borovinskaya (Head of the work). We are lucky that we could find an investor for this work.

8.5. THE MAIN RESULT

As a result a boundary area between the combustion theory and materials science has been created and called Structural Macrokinetics. It has brought combustion and materials science together: the specialists in combustion became more interested in materials science and those in materials science got rid of their fear of combustion and accepted its creative role. A bridge between the combustion practice and materials technology has been made. New terms –"material-generating combustion processes" and "combustion product materials science"- have appeared.

The main results of our studies can be summarized as follows.

1. New subjects of inquiry and techniques for use in chemistry, physics, and mechanics have been proposed.

2. New phenomena, effects, events, concepts, and processes have been studied and described. New theories and lines of research in combustion science, non-linear dynamics, macroscopic kinetics, and materials science have been suggested. As a result, a new scientific discipline, structural Macrokinetics, has appeared.

3. Generation of weak electric/magnetic fields as well as of electromagnetic and acoustic oscillations in reacting media has been established, which opened up new horizons for investigating the mechanism of chemical reactions in combustion waves.

4. It was suggested to regulate self-propagating chemical reactions by means of external influences and inert additives.

5. The method of self-propagating high-temperature synthesis (SHS) has been suggested as a novel technique for production of materials and compounds. A huge number of inorganic compounds and materials have been synthesized.

6. A concept of the alternative technology for optimal utilization of reaction heat has been formulated. New highly efficient SHS-based production processes have been developed.

7. High profitability of technology-intensive SHS facilities has been proved.

Favorable Situation

The World SHS community has been created. It united the specialists whose concentrated efforts provided the further consistent development of this scientific-and-technical direction.

Modern Definition of SHS

Nowadays SHS is a high-temperature process occurring at self-sustaining modes such as:

-   Autowave (or self-propagating) mode,

-   Induction (or accelerating) one,

-   Stabilized in space,

-   And in material-forming media (solid, liquid, gaseous, and multiphase ones) yielding

-   Powders of various particle shapes and sizes, including nano-sized ones,

-   Porous materials and items including those with nano-sized pores,

-   Compact non-porous materials (sintered, compacted by external forces, cast and also those with nano-sized grains),

-   Films and coatings of different thickness including nano-sized ones.

The products are characterized by various chemical, physical and mechanical properties, they can be produced under industrial terms by the science-intensive technology which is alternative to a conventional one, and used in scientific investigations, medicine and various industries [68 – 81].

40 Years of SHS: A Lucky Star of a Scientific Discovery, 2012, 90-97

CHAPTER 9

Miscellaneous

Abstract: Various problems and events accompanying the SHS development are considered as well as the SHS infrastructure is described.

9.1. ADJOINING AREAS AND RESULTS

In this chapter of my book I'd like to present some arguments about SHS.

First of all, science and technology had existed before the SHS appeared. SHS "was born" in the investigation of condensed system combustion. So, combustion theory and practice influenced its development greatly. We used some methods of experimental investigations of combustion regularities, measurement of combustion temperature and velocity, determination of temperature profiles, methods of thermodynamic calculations, and basic equations of flame propagation theory.

SHS also borrowed much from materials science and technology. Materials science has enriched us with the knowledge of materials, their production methods, and application fields. We used some available methods for analyzing synthesized materials and substances, preparing green mixtures.

Chemical physics (alma mater of modern combustion science) has provided us with the concepts of reaction kinetics in heterogeneous media at high temperatures, analysis methods of chemical reactions in combustion wave. We have also got useful information from high pressure technique. Our knowledge in the theory of heat- and mass-transfer helped us a lot.

SHS says to all these scientific areas: "Thank you ever so much, my dear colleagues!"

How has SHS repaid? Our effort to answer this question and show you the contribution of SHS to the development of adjoining areas is shown in Fig. **9.1.1**. I'd like to underline once more that we had the main direction in the development of SHS – we wanted to found a new field of knowledge – a boundary area between combustion theory and materials science, to develop scientific, technological and materials-science backgrounds of this field, and to find application fields for our achievements. We have been bound for this direction.

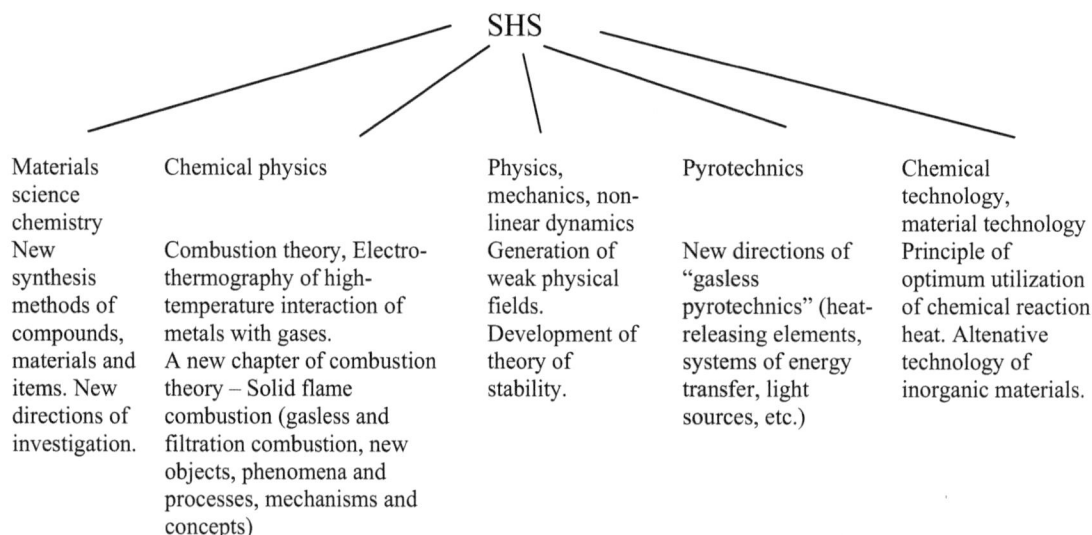

SHS

Materials science chemistry	Chemical physics	Physics, mechanics, non-linear dynamics	Pyrotechnics	Chemical technology, material technology
New synthesis methods of compounds, materials and items. New directions of investigation.	Combustion theory, Electro-thermography of high-temperature interaction of metals with gases. A new chapter of combustion theory – Solid flame combustion (gasless and filtration combustion, new objects, phenomena and processes, mechanisms and concepts)	Generation of weak physical fields. Development of theory of stability.	New directions of "gasless pyrotechnics" (heat-releasing elements, systems of energy transfer, light sources, etc.)	Principle of optimum utilization of chemical reaction heat. Altenative technology of inorganic materials.

Fig. (9.1.1): Contribution of SHS to adjacent fields of science and technology.

Certainly, there were some incidental results which were rather attractive for us. I'd like to mention two of them. From my point of view they are the most significant.

- Discovery of thermal auto-oscillations of the combustion front and appearance of spin waves leading to the development of our concept on the thermal nature of autowave process instability. This discovery relates to nonlinear dynamics and is of great interest for specialists.

- Generation of weak physical fields in SHS processes. It proves the idea of a complicated mechanism of elementary reactions of SHS. I think it should be of a particular interest for specialists in physics and electrical chemistry. But I am sure that they have not heard about our results in spite of the fact that more than 50 articles have been published.

Year	SHS system, reaction	Probe configuration	Field parameter	Authors
1980-1981	Mo-B	2 thermal electrodes along the sample	~thermal EMF	Levakov E.V., Peleskov S.A., Sorokin V.P.
1990-1994	$C_3H_4O_4$-$C_4H_{10}O_2$	Matrix of thermal electrodes in the front plane	0.7 V	Merzhanov A.G., Shulman Z.P., Khusid B.M., Dragun V.L., Klimchuk E.G., Mansurov V.A., Ovchinnikova S.M., Ubortsev A.D.
1996-2007	Nb-B, Ta-C, Mo-B, Ti-N(Si,B), Ni-Al, Fe-Al(Ti), Ba-Cr-O, Li-Fe-O, Na-Cu-O, Y-Ba-NaClO$_4$-Cu, Al-Cr-Mg-O, Ti-Cr-O, Na-Fe(Nb)-O, Cr-O, Cr-B-O, Ba-Fe-NaClO$_4$, LiIO$_3$-Cr(B), NaClO$_4$-Cr(B), KClO$_4$-Cr(B,Fe), Mg(ClO$_4$)$_2$-B(Cr), Mg-Ti-O, Ca-Ti(Cr,B)-O, Sr-Cr(B)-O, W-Ba-O, Ba(ClO$_4$)$_2$-Fe(Cr), Mo-Ba-O	←	0.03 -1.7 V	Morozov Yu.G., Kuznetsov M.V., Nersesyan M.D., Merzhanov A.G., Bakhtamov S.G., Busurin S.M., Chernega M.L., Chobko A.A.
1996-2001	Ti(air), Ti-N(C,Si), Mo-Si	L ←	0.3 A/mm^2, 2 mA	KudryashovV.A. Mukasyan A.S., Filimonov I.A., Kamynina O.K., Kidin N.I., Rogachev A.S., Umarov L.M.
1999-2006	Zr-WO$_3$, Al-Ni, Ti-C		2 mV – 0.8 V	Proskudin V.F.
2000-2004	Co-S, Ni-Al	L L	4–6 V	Maksimov Yu. M., Kirdyashkin A.I., Korogodov V.S., Polyakov V.L.
2001-2006	Cr-O, Co-O, Fe-O, Nb-O, Mg-O, Ti-O, Li-Fe(Mn)-O, Na-Fe-O, Ba-Fe(Ti)-O, Sr-Fe-O, Ti-Zr-O, Mn-O, Ni-O, Mg-O, Al-O, Fe-O-NaClO$_4$, Ti-O-NaClO$_4$, Al-Ni, SrCO$_3$-Fe, PbO-Fe, Ti-N, Pb(NO$_3$)$_2$-Fe	←	0.05-1.7 V	Nersesyan M.D., Claycomb J.R., Ritchie J.T., Miller J.H., Luss D., Richardson J.T., Martirosyan K.S., Filimonov I.A., Setoodeh M.
2003	Zr-O, Ti-O, Fe-O, Ni-O	L•J	0.25-2 V, 100 mA	Martirosyan K.S., Filimonov I.A., Nersesyan M.D., Luss D.

Fig. (9.1.2): Generation of weak electric fields in SHS.

Fig. **9.1.2** and **9.1.3** demonstrate the most important results achieved in physical field generation.

Measurement of SHS process characteristics (combustion wave temperature and velocity, impuritive gas volume, thermal effect, etc) proves that SHS processes are also interesting for specialists in pyrotechnics.

I can continue enumerating. But I think it's not the main point.

Years	SHS-system, product	Field characteristics	Field parameters	Authors
1992-1996	$SrFe_{12}O_{19}$	Residual field in the products after synthesis	20 μT	Merzhanov A.G., Mktrchan S.O., Nersesyan M.D., Avakyan P.B.
1998	Li-Fe-O	During synthesis, inside magnetometer	Relative measurements	Morozov Yu.G.
1999-2004	Ba-Ti-O, Sr-Ti(Si)-O, Ba-Cr-O, Ti-O-$NaClO_4$, Fe-Al-O, Na-Fe--O, Cr-O, Co-O, Fe-O, Nb-O, Mg-O, Li-Fe-O, Sr-Fe-O, Ba-Fe-O	During synthesis, outside the sample; outside the sample /after synthesis/ after cooling	0.6-6.5/8/800 nT	Nersesyan M.D., Claycomb J.R., Ritchie J.T., Miller J.H., Richardson J.T., Luss D.
2001	Zr-O, Hf-O, Ti-O, Fe-O, Mn-O	During synthesis, outside the sample	7 – 13 nT	Nersesyan M.D., Luss D.,Claycomb J.R., Ritchie J.T., Miller J.H.
2003	$SrFe_{12}O_{19}$, $BaFe_{12}O_{19}$, $CoFeO_4$, $Li_{0.5}Fe_{2.5}O_2$, $Yi_3Fe_5O_{12}$, $PbFe_{12}O_{19}$	After cooling, Outside the sample	4 – 8 μT	Martirosyan K.S., Claycomb J.R., Gogoshin G., Yarbrough R.A., Miller J.H., Luss D.

Fig. (9.1.3): Generation of weak magnetic fields in SHS.

9.2. WHAT COULD WE HAVE DONE?

We can find the answer to this question in Fig. **9.2.1**.

Development of continuous technological processes is of great importance. In the chemical technology this problem has been successfully solved. In combustion practice such processes have been organized for gases and gaseous systems. But in our field the matter concerns continuous combustion processes of condensed systems (i.e. gasless and filtration combustion). It is difficult to organize them. But there are some attractive proposals (Fig. **9.2.2**), and perhaps, they can provide the technological future of SHS.

- To develop combustion models corresponding to experimental conditions and use them for comparison of experimental and calculation data obtained for various systems;
- To organize purposeful investigation of mechanism and kinetics of high-temperature interaction in combustion waves;
- To develop technologies based on thermally conjugate SHS processes (simultaneous production of two or more products);
- To organize continuous SHS processes and technologies with transformation of thermal energy to electric power;
- To automate forced SHS-compaction technology;
- To apply direct SHS production technologies wider;
- To develop and apply a set of technological approaches with strong energy and mechanic effects;

- To develop special SHS equipment combining various operations;
- To develop technologies of direct production of inorganic materials strengthened by nano-sized grains;
- To build a pilot plant of SHS powder metallurgy.

Fig. 9.2.1: SHS: what else can be done?

The problem of SHS heat utilization is closely connected with the development of continuous technological processes. One of the possible ways is transformation of thermal energy to electric power. In this case we would be able to get a required product and electric power simultaneously, i.e. we could organize a plant and an electric power station at the same place, though the electric station would be used for local purposes.

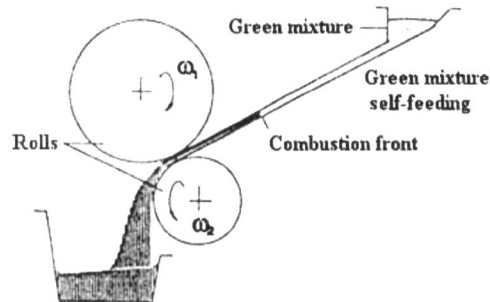

Rolling reactor

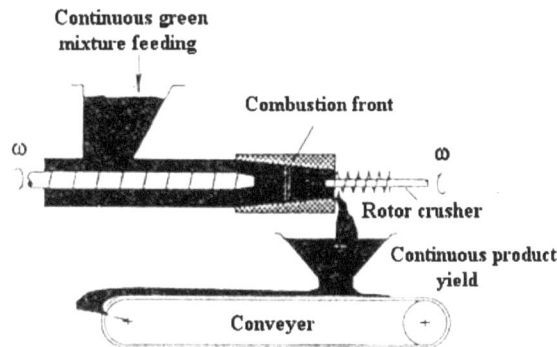

Conveyer-type reactor

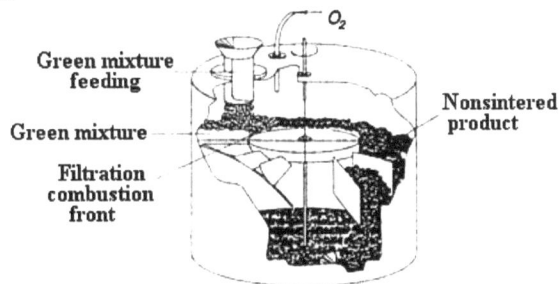

Rotor-type reactor

Fig. (9.2.2): Some ideas of continuous SHS process organization.

There are some ways of direct utilization of chemical reaction heat. The easiest one is common organization of high- and low-exothermic processes. It is used for heating premises and called a chemical furnace; I call it "thermally conjugate processes". One can think of many variations of such a process application in order to achieve an efficient heat exchange and easy separation of the synthesized products, to get rid of the product contamination.

When we invented the process we thought that it would be widely used soon. But we were wrong. I can't understand the reason. Nevertheless, I would like to find a profitable application for this method.

9.3. ON THE FIRST INTERNATIONAL MEETING WITH SPECIALISTS IN POWDER METALLURGY

My report at the meeting was of fundamental importance. At that time our Russian specialist in metallurgy, especially in powder metallurgy, did not recognize SHS. I worked hard at my report, gave a lot of examples with some figures and comparison. The title of the talk was "Self-propagating High-temperature Synthesis and Powder Metallurgy: Unity of Goals and Competition of Principles". There were two main parts: "SHS-products as raw materials for powder metallurgy" and "SHS method as an alternative to conventional methods of powder technology". This report became famous and played an important role in establishing contacts with specialists in materials science and metallurgy (Fig. **9.3.1**).

Powder metallurgy

Stage 1, powder production

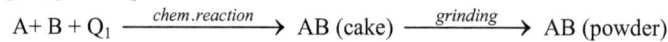

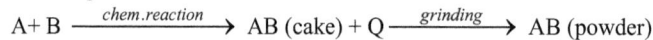

$$A + B + Q_1 \xrightarrow{chem.reaction} AB\ (cake) \xrightarrow{grinding} AB\ (powder)$$

Stage 2, material production

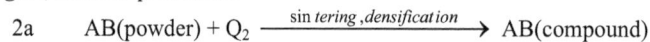

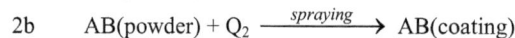

2a $AB(powder) + Q_2 \xrightarrow{sintering, densification} AB(compound)$

2b $AB(powder) + Q_2 \xrightarrow{spraying} AB(coating)$

Self-propagating high-temperature synthesis

Scheme 1.
Powder production

$$A + B \xrightarrow{chem.reaction} AB\ (cake) + Q \xrightarrow{grinding} AB\ (powder)$$

Stage 2. Material (item) production
See **2a** and **2b**

Scheme 2.
Production of material (item) with preset shape

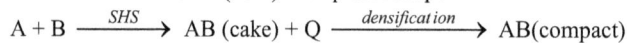

$$A + B \xrightarrow{SHS} AB\ (cake) + Q \xrightarrow{densification} AB(compact)$$

Other schemes

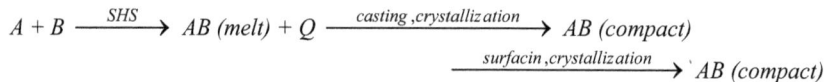

$$A + B \xrightarrow{SHS} AB\ (melt) + Q \xrightarrow{casting, crystallization} AB\ (compact)$$

$$\xrightarrow{surfacin, crystallization} AB\ (compact)$$

$$A + B \xrightarrow{SHS} AB\ (powder) + Q \qquad SHS\ in\ gas\ or\ gas\ suspension$$

$$A + B + additions \xrightarrow{SHS} AB\ (coating) + Q$$

Fig. (9.3.1): Comparison of SHS and Powder Metallurgy technologies.

9.4. "COMMUNICATIVE" SHS

There are a lot of ways for SHS future development. I'd like to tell you about one of them.

SHS is easily combined with any conventional processes due to its easy organization. It is not difficult to develop a joint process and there are a lot of examples. In any case we must use a green mixture but not a material to be processed. The material is formed during the process.

Below you can see an example

SHS +

Heating (in a furnace, Joule, in a microwave)

Forced compaction (from slow isostatic heating to fast shock-wave heating)

Sintering

Surfacing

Spraying

etc.

It is possible to discover a lot of original solutions and unexpected results, and I'd like to underline that due to the process flexibility I called it "communicative".

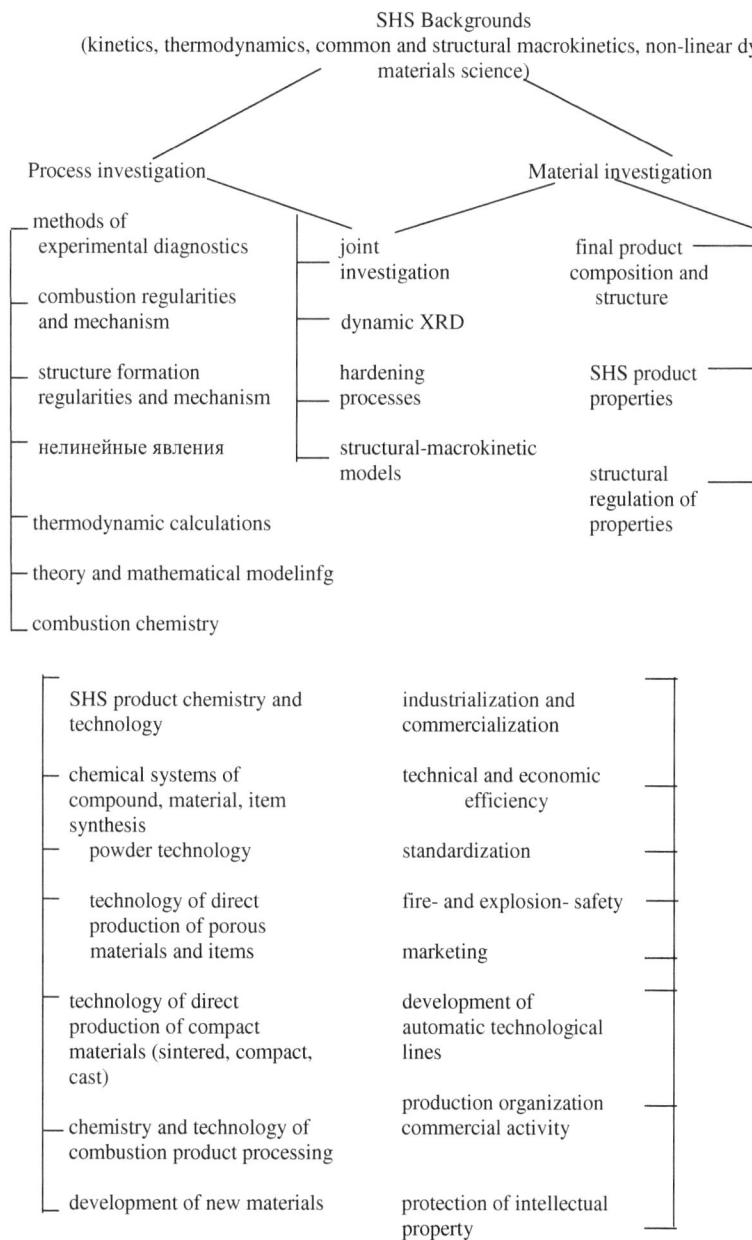

SHS Backgrounds
(kinetics, thermodynamics, common and structural macrokinetics, non-linear dynamics,
materials science)

Process investigation   Material investigation

- methods of experimental diagnostics
- combustion regularities and mechanism
- structure formation regularities and mechanism
- нелинейные явления
- thermodynamic calculations
- theory and mathematical modelinfg
- combustion chemistry

- joint investigation
- dynamic XRD
- hardening processes
- structural-macrokinetic models

- final product composition and structure
- SHS product properties
- structural regulation of properties

- SHS product chemistry and technology
- chemical systems of compound, material, item synthesis
  - powder technology
- technology of direct production of porous materials and items
- technology of direct production of compact materials (sintered, compact, cast)
- chemistry and technology of combustion product processing
- development of new materials

- industrialization and commercialization
- technical and economic efficiency
- standardization
- fire- and explosion- safety
- marketing
- development of automatic technological lines
- production organization commercial activity
- protection of intellectual property

Fig. (9.5.1): Infrastructure of self-propagating high-temperature synthesis.

9.5. INFRASTRUCTURE

Nowadays SHS is a rather successful area with well-developed infrastructure (Fig. **9.5.1**). It consists of 4 blocks:

1. Investigation of the process.

2. Investigation of materials. Also common investigation of processes and products.

3. Development of technologies.

Organization of science-intensive production facilities.

9.6. A LUCKY STAR OF A SCIENTIFIC DISCOVERY

I have been already asked about the reason of such a title of my work. On the face of it, it is incomprehensible. But I have explained.

At the beginning SHS was being developed rather slowly but with self-acceleration. It is very important because it means that in each subsequent period SHS becomes stronger and stronger; the number of people working in the area, the quantity of articles and patents, production lines and facilities are constantly increasing. Parents do not usually notice their child growing because they can see him every day; and only then they can realize that he is already adult.

It has happened to us. I was searching some information in the I-net and suddenly came across a note about our process. Then I decided to find the information about SHS and was impressed by its abundance. There were a lot of data in various information systems – Google, Yandex, Rambler, etc. I was searching using three key expressions: SHS, self-propagating high-temperature synthesis and combustion synthesis. At first, I was disappointed: I used the abbreviation SHS and found out that it was a very popular one. But in the glossary I have found only one explanation – self-propagating high-temperature synthesis. It means that our SHS is the most important one. The other words were connected with our work, and it was possible to find any information about it. For example, there were about 1,200,000 sites about combustion synthesis in spring but in autumn their number became ~19,500,000. One can see texts of articles, abstracts, proceedings of conferences, reports, commercial information and so on. After getting acquainted with all the sites, one can become a great specialist in SHS. But the most impressive for me were new directions of our process. It appeared that I was not aware of much and was far behind the times. I'd like to give an example of a curious thing: SHS is used for destruction of confidential information. Besides, in Institutes and Universities they hold contests of the best theses in SHS.

I thought a lot about it and developed a conception which I called later "Four stages in the "life" of a lucky scientific discovery". Many discoveries can't be realized and they are easily forgotten without becoming famous. Others flare up rather notably causing a sensation among specialists, but they go out very quickly too. We can't say that those discoveries are lucky.

In our case the situation was different. We developed our direction rather slowly, we argued with our opponents trying to prove that we were right, but we continued advancing. When I use the pronoun "we", I mean our friendly SHS community, all the specialists in SHS all over the world.

Below I'd like to describe these four stages:

Stage 1. None except the authors of the discovery is engaged in the work. The authors fulfill their task rather quietly and slowly.

Stage 2. Scientists from other research centers become interested in the discovery and begin working with the authors. The authors know all the specialists engaged in the work.

Stage 3. The problem is being developed with self-acceleration. The number of scientists and their works is so huge that the authors can't follow and influence the process R&D.

Stage 4. The problem has overstepped the limits of induction and begun developing very fast. Beginners are not interested in the problem's history or do not know the pioneers' names.

I think, nowadays we observe the beginning of Stage 4 of the lucky star of our scientific discovery.

By the way, I don't feel hurt when my name is not referred. It seems to me that I am a composer and watch the people who like my songs but think that they are folk heritage.

That is why I called my paper "SHS: a lucky star of a scientific discovery".

REFERENCES

[1]   Merzhanov A.G., Shkiro V.M., Borovinskaya I.P. Synthesis of Refractory Inorganic Compounds USSR Inventor's Certificate 255 221, **1967**; Byull. Izobr., **1971**, No.10; Fr. Patent 2 088 668, **1972**; US Patent 3726643, **1973**; UK Patent 1 321 084; Jpn. Patent 1 098 839, **1982**.

[2]   Merzhanov A.G., Borovinskaya I.P. Self-Propagating High-Temperature Synthesis of Inorganic Compounds, *Dokl. Akad. Nauk SSSR*, **1972**, vol. 204, No. 2, pp. 366-369.

[3]   Merzhanov A.G., Borovinskaya I.P. A new class of combustion processes, *Combust Sci Tecnol,* **1975**, vol. 10, No. 5-6, pp. 195-200.

[4]   Merzhanov A.G. Twenty years of search and findings, *Combustion and plasma synthesis of high-temperature materials,* Eds. Z.A. Munir, J.B. Holt. N.Y.: VCH Publ. Inc., **1990**, pp. 1-53

[5]   Crider J.F. Self-propagating high-temperature synthesis – a soviet method for producing ceramic materials, *Ceram. Eng. Sci. Proc.,* **1982**, vol. 3, No. 9-10, pp. 519-528.

[6]   Merzhanov A.G. Solid Flames: Discovery, Concepts, and Horizons of Cognition, *Combust. Sci. Technol.,* **1994**, vol. 98, No. 4-6, pp. 307-336.

[7]   Merzhanov A.G., Borovinskaya I.pp, Volodin Yu.E. On combustion mechanism of porous metal samples in nitrogen, *Dokl. Akad. Nauk SSSR*, **1972**, vol. 206, No. 4, pp. 905-908.

[8]   Merzhanov A.G., Filonenko A.K., Borovinskaya I.P. New Phenomena in Combustion of Condensed Systems, *Dokl. Akad. Nauk SSSR*, **1973**, vol. 208, No. 4, pp. 892-894.

[9]   Azatyan T.S., Maltsev V.M., Seleznev V.A. On mechanism of combustion wave propagation in titanium-boron mixtures, *Fiz. Goreniya Vzryva*, **1980**, No. 2, pp. 37-42.

[10]   Dolukhanyan S.K., Sarkisyan A.R., Pogosyan A.S. SHS-formed molybdenum disilicide in the high-temperature heaters production, *Promyshl. Armenii (Erevan),* **1976**, No.1, pp. 46-47.

[11]   Dolukhanyan S.K., Nersesyan M.D., Nalbandyan A.B., Borovinskaya I.P., Merzhanbov A.G. Transition metal combustion in hydrogen, *Dokl. Akad. Nauk SSSR,* **1976**, v. 231, No. 3, pp. 675-678.

[12]   Martirosyan N.A., Dolukhanyan S.K., Mkrtchyan G.M., Borovinskaya I.P., Merzhanov A.G. Purification Processes during SHS of Refractory Compounds, *Poroshk. Metall.,* **1977**, No. 7, pp. 36-40.

[13]   Merzhanov A.G., Grigor'ev Yu.M., Kharatyan S.L., Mashkinov L.B., Vartanyan Zh.S. Study on Heat-Evolution Kinetics at High-Temperature Nitration of Zirconium Wires, *Fiz. Goreniya Vzryva*, **1975**, vol. 11, No. 4, pp. 563-568.

[14]   Kharatyan S.L., Sardaryan Yu.S., Sarkisyan A.A., Merzhanov A.G. Carbides Formation during High-Temperature Interaction of Zirconium and Tantalum with Simple Hydrocarbons, *Problemy tekhnologicheskogo goreniya* (Problems of Technological Combustion), Chernogolovka: Izd. Inst. *Chem. Phys.,* **1981**, vol. 2, pp. 37-40.

[15]   Kharatyan S.L., Manukyan Kh.V., Nersisyan H.H., Khachatryan H.L. Macrokinetic laws of activated combustion at synthesis of composite ceramic powders based on silicon nitride, *Int. J. SHS,* **2003**, vol. 12, No. 1, pp. 19-34.

[16]   Maksimov Yu.M., Ziatdinov M.Kh., Merzhanov A.G., Raskolenko L.G., Lepakova O.K. Combustion of vanadium-iron alloys in nitrogen, *Fiz. Goreniya Vzryva*, **1984**, vol. 20, No. 5, pp.16-21. .

[17]   Smolyakov V.K., Maksimov Yu.M. Structural transformation of powder media in the wave of self-propagating high-temperature synthesis, *Int. J. SHS.* **1999**, vol. 8, No. 2, pp. 221-249.

[18]   Maksimov Yu.M., Kirdyashkin A.I., Korogodov V.S., Polyakov V.L. Generation and transfer of electric charge in self-propagating high-temperature synthesis using Co-S, *Combust. Explosion Shock Waves,* **2000**, vol. 36, No. 5, pp. 670-673.

[19]   Maksimov Yu.M., Itin V.I., Smolyakov V.K., Kirdyashkin A.I. SHS in electric and magnetic field, *Int. J. SHS,* **2001**, vol. 10, No. 3, pp. 259-329.

[20]   Smolyakov V.K., Lapshin O.V., Maksimov Yu.M. Nonisothermal interaction of powders with a reactive gaseous medium during grinding, *Combust. Explosion Shock Waves,* **2003**, vol. 39, No. 6, pp. 659-669.

[21]   Lepakova O.K., Raskolenko L.G., Maksimov Yu.M. Self-propagating high-temperature synthesis of composite material TiB_2-Fe, *J. Mater. Sci.,* **2004**, vol.39, No. 11, pp. 3723-3732.

[22]   Merzhanov A.G., Karyuk G.G., Borovinskaya I.P., Sharivker S.Yu., Moshkovskii E.I., Prokudina V.K., Dyadko E.G. Titanium carbide produced by SHS – a highly efficient abrasive material, *Poroshk. Metall.,* **1981**, No. 10, pp. 50-59.

[23] Molodovskaya E.K., Petrunin V.F., Karimov I., Kalanov M., Khaidarov G., Borovinskaya I.P., Pityulin A.N, Merzhanov A.G. Neutronographic Study of Cubic Tantalum Nitrides, *Fiz. Met. Metalloved.*, **1975**, vol. 40, No. 1, pp. 202-204.

[24] Em V.T., Karimov I., Petrunin V.P., Khidirov I., Latergays I.K., Merzhanov A.G., Borovinskaya I.P., Prokudina V.K. Neutronographic Investigation of Ordering in Titanium Carbides, *Kristallographiya*, **1975**, vol. 20, No. 2, pp. 320-323.

[25] Kvanin V.I., Gorovoi V.A., Balikhina N.T., Borovinskaya I.P., Merzhanov A.G. Investigation of the processes of forced SHS compaction of large-scale hard-alloy articles, *Int. J. SHS*, **1993**, vol.2, No.1, pp. 56-68.

[26] Kvanin V.L., Balikhina N.T., Vadchenko S.G. Combustion of hollow cylinders, *Combust. Explosion Shock Waves*, **2002**, vol. 38, No. 4, pp. 425-429.

[27] Kvanin V.L., Balikhina N.T., Borovinskaya I.P., Shujun Li, Di Cao, Development of forced SHS compaction to produce a construction alloy with improved mechanical properties in the Ti-C-Ni-Mo system, *Int. J. SHS*, **2004**, vol. 13, No. 2, pp.161-169.

[28] Merzhanov, A.G. Fluid dynamics phenomena in the processes of self-propagating high-temperature synthesis. Ceram. Int., 1997, 5(2), 119-163.

[29] Merzhanov A.G., Borovinskaya I.P., Nersesyan M.D., Peresada A.G., Morozov Yu.G. Self-propagating high-temperature synthesis of high-temperature super-conductors, *Dokl. Akad. Nauk SSSR*, **1990**, vol. 311, No. 1, pp. 96-101.

[30] Merzhanov A.G., Self-propagating high-temperature synthesis of ceramic (oxide) superconductors, *In: Ceramic Transactions. Superconductivity and Ceramic Superconductors/ Eds. K.N. Nair, E.I. Dupont de Nemours, E.A. Giess. Westerville, Ohio: Amer. Ceram. Soc. Publ.*, **1990**, vol. 13, pp.519-549.

[31] Merzhanov A.G., Rogachev A.S., Structural macrokinetics of SHS processes, *Pure Appl. Chem.*, **1992**, vol.64, No. 7, pp. 941-953.

[32] Grigoryan H.E., Rogachev A.S., Sytschev A.E. Gasless combustion in the Ti-Si-C system, *Int. J. SHS*, **1997**, vol.5, No.1, pp. 29-39.

[33] Rogachev A.S. Macrokinetics of gasless combustion: Old problems and new approaches, *Int. J. SHS*, **1997**, vol.6, No.26, pp. 215-242.

[34] Ponomarev V.I., Khomenko I.O., Merzhanov A.G. Experimental Method of Time-Resolved X-Ray Diffraction, *Kristallografiya*, **1995**, vol. 40, No. 1, pp. 14-17.

[35] Kovalev D.Yu., Shkiro V.M., Ponomarev V.I. Dynamics of Phase Formation during Combustion of Zr and Hf in Air, *Int. J. SHS*, **2007**, vol. 16, No. 4, pp. 169-175.

[36] Merzhanov A.G., Bloshenko V.N., Bokii V.A., Borovinskaya I.P., Efimov O.Yu., Sharivker S.Yu. Porous SHS materials on titanium carbide base. *Dokl. Akad. Nauk SSSR*, **1992**, vol. 324, No. 5, pp. 1046-1050.

[37] Gladun G.G., Chernoglazova T.V., Ksandopulo G.I. Non-stationary films combustion on the support. *In: Proc. Of the Russ.-Jap. Seminar on Combustion, Chernogolovka, 2-5 Oct. 1993, Moscow: Russ.Section Combust. Inst. Publ.*, **1993**, pp. 147-148.

[38] Tavadze, G.F., Natsvlishvili T.N., Bezhitadze D.T. Phase formation in the Ti-B system, Soobshch. 2 Akad. Nauk Gruzii, **1994**, No.5.

[39] Tavadze G.F., Natsvlishvili T.N., Bezhitadze D.T. Heat-resistant alloy STIM-4, Proceedings of the I Int. Symp. On SHS, Alma-Ata, **1991**, pp. 35.

[40] Oniashvili G.Sh., Yukhvid V.I. Centrifugal SHS fusion in steel, Liteinoe Proiz-vo (Moscow), **2001,** (7).

[41] Oniashvili G.Sh., Gedevanishvili Sh.V., Yukhvid V.I. Features of chemical conversion of highly caloric mixtures on the basis of ore concentrate in the combustion mode. Inzh.-Fiz. Zh., **1993,** 65(5).

[42] Amosov A.P., Bichurov G.V., Bolshova N.F., Erin V.M., Makarenko A.G., Markov Yu.M. Azides as reagents in SHS processes, *Int. J. SHS*, **1992**, vol. 1, No.2, pp. 239-245.

[43] Amosov A.P., Makarenko A.G., Samboruk A.R., Seplyarskii B.S., Skobeltsov V.P., Zakamov D.V. SHS filtration combustion technique of ceramic powders, *Int. J. SHS*, **1998**, vol. 7, No. 4, pp. 423-438.

[44] Levashov E.A., Rogachev A.S., Borovinskaya I.P., Yukhvid V.I. Physico-chemical and *Technological Fundamentals of Self-propagating High-temperature Synthesis*, Binom, Moscow **1999**.

[45] Levashov E.A., Padyukov K.L. Self-propagating high-temperature synthesis: a new method for production of diamond-containing materials. *Diam Relat Mater,*1993, vol.2, No.2, pp. 207-210.

[46] Borovinskaya I.P. Chemical Classes of SHS Processes and Materials, *Pure Appl. Chem.*, **1992**, vol. 64, No. 7, pp. 919-940.

[47] Borovinskaya I.P. The Routes of Self-Propagating High-Temperature Synthesis, Proceedings of PAC RIM, **1993**, Honolulu, pp. 159.

[48] Merzhanov A.G. Self-Propagating High-Temperature Synthesis and Powder Metallurgy: Unity of Goals and Competition of Principles, in Particulate Materials and Processes. *Advances in Powder Metallurgy*, vol. 9, Princeton: Metal Powder Ind. Fed. Publ., **1992**, pp. 341-368.

[49] Yukvid V.I. Modifications of SHS Processes, Pure Appl. Chem., 1992, vol.64, No.7, pp. 977-988.

[50] Yukvid V.I., Vishnyakova G.A., Silyakov S.L., Sanin V.N., Kachin A.R. Structural macrokinetics of alumothermic SHS processes, *Int. J. SHS*, **1996**, vol.5, No. 1, pp. 93-105.

[51] Rogachev A.S. Macroheterogeneous mechanism of gasless combustion, *Fiz. Gorenia Vzryva*, **2003**, vol. 39, No. 2, pp. 38-47.

[52] Borovinskaya I.P., Loryan V.E. SHS of Titanium Nitrides at High Nitrogen Pressures, *Poroshk.. Metall.*, **1978**, No. 11, pp. 42-45.

[53] Borovinskaya I.P. Formation of Refractory Compounds during Combustion of Heterogeneous Condensed Systems, Proceedings of the IV All-Union Symp. on Combustion and Explosion, Moscow: Nauka, **1977**, pp. 138-148.

[54] Merzhanov A.G., Kovalev D.Yu., Shkiro V.M., Ponomarev V.I. Equilibrium Products of Self-Propagating High-Temperature Synthesis, *Dokl. Phys. Chem.,* **2004**, vol. 394, No. 2, pp. 34-38.

[55] Molodovskaya E.K., Petrunin V.F., Karimov I., Kalanov M., Khaidarov G., Borovinskaya I.P., Pityulin A.N, Merzhanov A.G. Neutronographic Study of Cubic Tantalum Nitrides, *Fiz. Met. Metalloved.*, **1975**, vol. 40, No. 1, pp. 202-204.

[56] Em V.T., Karimov I., Petrunin V.P., Khidirov I., Latergays I.K., Merzhanov A.G., Borovinskaya I.P., Prokudina V.K. Neutronographic Investigation of Ordering in Titanium Carbides, *Kristallographiya,* **1975**, vol. 20, No. 2, pp. 320-323.

[57] Ivleva T.P., Merzhanov A.G. Three-dimensional modeling of solid-flame chaos, *Dokl. Phys. Chem.*, **2003**, vol. 391, No. 1-3, pp. 171-173.

[58] Grachev V.V., Ivleva T.P., Borovinskaya I.P. Filtration combustion in an SHS reactor, *Int. J. SHS*, **1995**, vol.4, No.3, pp. 245-252.

[59] Grachev V.V., Ivleva T.P. Surface and layer-by-layer combustion modes in gas-solid systems, *Int. J. SHS*, **1998**, vol.7, No.1, pp.1-19.

[60] Holt J.B., Wong J., Larson E.M., Waide P.A., Rupp B., and Frahm B. Time-Resolved X-Ray Diffraction Study of Solid Combustion Reactions, *Science*, **1990**, vol. 249, pp. 1406-1409.

[61] Larson E.M., Wong J., , Holt J.B.,. Waide P.A, Nutt G., Rupp B. Termninello L.J., Time-Resolved Diffraction Study of Ta — C Solid Combustion System, *J. Mater. Res.*, **1993**, vol. 8, No. 7, pp. 1533-1541.

[62] Munir Z.A. Electrically stimulated SHS, *Int. J. SHS*, **1997**, vol.6, pp.165-185.

[63] Woolman J.N., Petrovic J.J., Munir Z.A. Microalloying of molybdenum disilicide with magnesium through mechanical and field activation, *J. Mater. Sci.*, **2004**.

[64] Aldushin A.P., Matkovsky B.J., Schult D.A. Buoyancy driven filtration combustion, *Comb. Sci. Technol.*, **1997**, vol. 125, pp. 283-349.

[65] La Salvia J.S., Meyers M.A., Kim D.K. Combustion synthesis/dynamic densification of TiC-Ni cermets. *J. Mater. Synth. Process.,* **1994**, vol. 2, No. 4, pp. 255-274.

[66] Puszynski J.A. Kinetics and Thermodynamics of SHS reactions, *Int. J. SHS*, **2001**, vol.10, No. 3, pp. 265-293.

[67] Puszynski J.A. Advances in the formation of metal and ceramic nanopowders, in Powder Materials: Current Research and Industrial Practices, *TMS Curr Res Indust Pract,* TMS, **2001**, pp. 89-105.

[68] Yi H.C., Guigne J.Y., Manerbino A.R., Robinson L.A., Ma J., Moore J.J. Modeling and fabrication of functionally graded materials by the combustion synthesis technique, *Mater. Sci. Forum*, **2003**, vol. 23, No. 425, pp. 239-244.

[69] Vandersall K.V., Thadhani N.N. Investigation of shock-induced and shock-assisted chemical reactions in Mo+2Si powder mixtures, *Metall. Trans. A*, **2003**, vol. 34, pp. 15-23.

[70] Nersesyan M.D., Claycomb J.R., Ritchie J.T., Miller J.H., Luss D., Magnetic fields produced by combustion of metals inh oxygen, *Combust. Sci. Technol.,* **2001**, vol.169, pp. 89.

[71] MartirosynK.S., Claycomb J.R., Gogoshin G., Yarbrough R.A., Miller J.H.Jr, Luss D. Spontaneous magnetization generated by spin, pulsating and planar combustion synthesis, *J. Appl. Phys.,***2003**, vol. 150, J9.

[72] Varma A, Mukasyan A.S., Deshpande K., Pranda P., Erii P. Combustion synthesis of nanoscale oxide powders : Mechanism, characterization and properties, *Mater. Res. Soc. Symp. Proc.*, **2003**, vol. 800, pp. 113-124.

[73] Rogachev A.S., Mukasyan A.S., Varma A. Thermal explosion modes in gasless heterogeneous systems, *j. Mater. Synth. Proc.,* **2002**, vol.10, No.1, pp. 29-34.

[74] Varma A., Rogachev A.S., Mukasyan A.S., Hwang S. Complex behavior of self-propagating reaction waves in heterogeneous media, *Proc. Natl. Acad. Sci. USA,* **1998**, No. 95, pp. 11053-11058.

[75] Mukasyan A.S., Marasia J.A., Filimonov I.A., Varma A. The role of infiltration on spin combustion in gas-solid systems, *Combust Flame,* **2000**, vol. 122, No. 3, pp. 368-374.

[76] Pelekh A., Mukasyan A.S., Varma A. Kinetics of rapid high-temperature reactions: Titanium-nitrogen system, *Ind. Eng. Chem. Res.,* **1999**, vol.38, No.4, pp. 563-568.

[77] Varma A., Mukasyan A.S., Deshpande K., Pranda P., Erii P. Combustion synthesis of nano-scale oxide powders : Mechanism, characterization and properties, *Mater. Res. Soc. Symp. Proc.,* **2003**, vol. 800, pp. 113-124.

[78] Odawara O. Long ceramic-lined pipes produced by a centrifugal-thermit process, *J. Am. Ceram. Soc.,* **1990**, vol.73, No. 3, pp. 629-633.

[79] Miyazaki E., Odawara O. Centrifugal effects on combustion synthesis of (Ti-B-C) compound system, *Mater. Res. Bull.,* **2003**, vol. 38, pp. 1375-1386.

[80] Miyamoto Y. Ecomaterials synthesis and recycling of nitriding combustion, *Solid State Mater. Sci.,* **2003**, vol. 7, pp. 241-245.

[81] Shibuya M., Ohyanagi M., Munir Z.A. Simultaneous synthesis and densification of titanium nitride/titanium diboride composites by high nitrogen pressure combustion, *J. Am. Ceram. Soc.,* **2002**, vol. 85, No. 12, pp. 2965-2970.

[82] Merzhanov A.G., Pityulin A.N. Self-Propagating High-Temperature Synthesis in Production of Functionally Gradient Materials, Proceedings of the III Int. Symp. on Structural and Functional Gradient Materials, **1994**, Lausanne, Ilschner B. and Cherradi N., Eds., Lausanne: Polytech. Univ. Romandes Publ., **1995**, pp. 87-94.

[83] Makino A. Range of flammability for the nonadiabatic SHS flame: theory and experimental comparisons. *Ann. Chim. Fr.,* **1995**, vol. 20, No. 3-4, pp. 139-150.

[84] Yuan R.Z., Fu Z.Y., Yang Z.L. Study on the preparation of TiB_2-$TiAl_3$-Al functionally graded materials by SHS method. *Proc. I Int. Symp. On FGM,* **1990**, Sendai, pp. 175.

[85] Yuan R.Z., Tu R., Zhang L. SHS of TiB_2-based multiphase ceramics and composites. *Int. J. SHS,* **2001**, *10*(4), 435.

[86] Ge C.C., Li J.T., Cao Y.G. Thermodynamics and combustion synthesis in Si-TiC-N and Si-TiO_2—N system, *Z. Metallkunde,* **1998**, vol. 89, No. 2, pp. 149-152.

[87] Zhang Shu Ge, Yan X., Sun G. Properties of ceramic-lined composite steel pipes and their application. *Int. J. SHS,* **2004**, *5*(1), 45-55.

[88] Zhang Shu Ge. The origin of SHS and its development in China, *Powder Metall. Technol.,* **1997**, vol. 15, No. 4, pp. 295-298.

[89] Lis J., Pampuch R., Piekarczyk J., Stobiierski L., Wisniewski A. Ceramic armor materials prepared by the self-propagating high-temperature synthesis, Proceedings of the 9th CIMTEC-World Ceramics Congress, Part 3, Ed. Vincenzini, P., Techna Srl., **1999**, pp. 269-275.

[90] Pampuch R., Stobierski L., Lis J. Use of SHS powders in synthesis of complex ceramic materials, *Int. J. SHS,* **2001**, vol. 10, pp. 201-214.

[91] Patil K.C., Aruna S.T., Mimani Tanu. Combustion synthesis: An update, *Cur. Opinion in Solid State and Mater. Sci.,* **2002**, No. 6, pp. 507-512.

[92] Gutmanas E.Y., Gotman I. Reactive synthesis of ceramic matrix composites under pressure, *Ceram. Int.,* **2000**, vol. 26, pp. 699-707.

[93] Orru R., Cincotti A., Concas A., Cao G. Development of processes for environmental protection based on self-propagating reactions, *Environm. Sci. Pollution Res.,* **2003,** vol. 6, pp. 385-389.

[94] Cao G., Doppiu S., Monagheddu M., Orru R., Sannia M., Cocco G. The thermal and mechanochemical self-propagating degradation of chloroorganic compounds: the case of hexachlorobenzene over calcium hydride, *Ind. Eng. Chem. Res.,* **1999**, vol.38, pp. 3218-3224.

[95] Munir Z.A., Anselmi-Tamburini U. Self-propagating exothermic reactions: the synthesis of high-temperature materials by combustion // *Mater. Sci. Rep.,* **1989**, vol. 3, No. 7-8, pp. 277-365

[96] Maglia F., Anselmi-Tamburini U., Deidda C., Delogu F., Cocco G., and Munir Z. A. Role of mechanical activation in SHS synthesis of TiC. *J Mater Sci,* **2004**, vol. 39, pp. 5227 – 5230.

[97]   Parkin I.P., Pankhurst Q.A., Affleck L., Aguas M.D., Kuznetsov M.V. Self-propagating high-temperature synthesis of $BaFe_{12}O_{19}$, $Mg_{0.5}Zn_{0.5}Fe_2O_4$ and $Li_{0.5}Fe_{2.5}O_4$: Time-resolved diffraction studies (TRXRD), *J. Mater. Chem.*, **2001**, vol.11, pp.193-199.

[98]   Kuznetsov M.V., Pankhurst Q.A., Parkin I.P., Affleck L., Morozov Yu.G. Self-propagating high-temperature synthesis of of chromium substituted lanthanum orthoferrites $LaFe_{1-x}Cr_xO_3$ ($0 \leq x \leq 1$), *J. Mater. Chem.*, **2001**, vol.11, pp. 854-858.

[99]   Agrafiotis C., Hlavacek V., Puszynski J.A. Direct synthesis of composites and solid solutions by combustion reactions. *Mater. Sci. Rep.,Combust. Sci. Technol.*, **1992**, vol. 88, pp. 187-199.

[100] Gladun G.G. (Xanthopoulou G.) Self-propagating high-temperature synthesis of catalysts and carriers. *Int. J. SHS*, **1994**, vol.1, No.3, pp.51-58.

[101] Bernard F., Paris S., Vrel D., Gailhanou M., Gachon J.C., Gaffet E. Time-Resolved XRD Experiments Adapted to SHS Reactions: Autoreview, *Int. J. SHS*, **2002**, vol. 11; *Int. J. SHS*, **2008**, vol. 17, No. 2, pp. 181-190.

[102] Bernard F., Gaffet E., Mechanical alloying in the SHS research, *Int. J. SHS*, **2001**, vol. 10, No. 2, pp. 109-132.

[103] Bernard F., Charlot F., Gaffet E., Munir Z.A., One-step synthesis and consolidation of nanoaluminides, *J. Am. Ceram. Soc.*, **2001**, vol. 84, No.5, pp. 10-914.

[104] Rogachev A.S., Gachon J.-C., Grigoryan H.E., Illekova E., Kochetov N.F., Nosyrev F.N.,. Sachkova N.V, Schuster J.C., Sharafutdinov M.K., Shkodich N.F., Tolochko B.P., Tsygankov P.A., Yagubova I.Y. Diffraction of Synchrotron Radiation for In-Situ Study of Heterogeneous Reaction Mechanisms in Lamellar Composites Obtained by Mechanical Activation and Magnetron Sputtering, *Nucl. Instrum. Methods Phys. Res.*, Sect. A, **2007**, vol. 575. pp. 126-129.

[105] Merzhanov A.G. Fundamentals, Achievements, and Perspectives for Development of Solid- Flame Combustion, *Russ. Chem. Bull.*, **1997**, vol. 46, No. 1, pp. 1-27.

[106] Merzhanov A.G. Self-Propagating High-Temperature Synthesis: Non-Equilibrium Processes and Equilibrium Products, *Adv. Sci. Technol.*, **2006**, vol. 45, pp. 36.

[107] Klimchuk E.G. Organic SHS, In: *SHS: Concepts of Current R@D,* Ed. Merzhanov A.G., Chernogolovka: Territoriya, **2003**, pp. 112-118 (in Russ.).

[108] Klimchuk E.G. Autowave exothermic organic synthesis in the mixes of organic solids, *Macromol Symp,* **2000**, vol. 160, pp. 107.

[109] Sytschev A.E., Merzhanov A.G. Self-Propagating High-Temperature Synthesis of Nanomaterials, *Russ. Chem. Rev.*, **2004**, vol. 73, No. 2, pp. 1479-1519.

[110] Borovinskaya I.P., Ignat'eva T.I., Vershinnikov V.I., Emelyanova O.M., Semenova V.N. SHS of Nano-sized Powders of Refractory Compounds, *Ross. Nanoteknnol.*, **2007**, vol. 2, pp. 114-119.

[111] Borovinskaya I.P., Ignat'eva T.I., Vershinnikov V.I., Emelyanova O.M., Semenova V.N. Self-Propagating High-Temperature Synthesis of Ultra- and Nano-sized Powder of Titanium Carbide, *Neorg. Mater.*, **2007**, vol. 43, pp. 1343-1350.

[112] Borovinskaya I.P., Ignat'eva T.I., Vershinnikov V.I., Sachkova N.V. Production of Nanosized Titanium Carbide Powder by SHS, *Neorg. Mater.,* **2004**, vol. 40, pp. 1190-1196.

[113] Borovinskaya I.P. SHS Ceramics: Synthesis, Technology, Application, *Nauka Proizvodstvu*, **2001**, No. 10 (48), pp. 11-18.

[114] Merzhanov A.G. *Advanced SHS Ceramics: Today and Tomorrow Morning in Ceramics: Toward the 21st Century*, N. Soga, A. Kato, Eds., Tokyo: Ceram. Soc. Jpn. Publ., **1991**, pp. 378-403.

[115] Borovinskaya I.P. SHS of Hard Alloys at the Threshold of the XXI Century, *Mashinostroitel*, **2000**, No. 3, pp. 15-20.

[116] Merzhanov A.G., Pityulin A.N. Self-Propagating High-Temperature Synthesis in Production of Functionally Gradient Materials, Proceedings of the III Int. Symp. on Structural and Functional Gradient Materials, **1994**, Lausanne, Ilschner B. and Cherradi N., Eds., Lausanne: Polytech. Univ. Romandes Publ., **1995**, pp. 87-94.

[117] Uvarov V.I., Borovinskaya I.P., Zagnitko A.V, Trotsenko N.M., Lukin E.S. Filters for Apparatuses for Production of Injection Solutions (Apyrogenic Water), *Ogneupory Tekhn. Keram.*, **2003**, No. 5, pp. 22-28.

[118] Merzhanov A.G. SHS on the Pathway to Industrialization. *Int. J. SHS*, **2001**, vol. 10, No. 2, pp. 237-257.

INDEX